Quantum Pythagoreans

Quantum Pythagoreans

Of Stars and Numbers, Gs and Waves

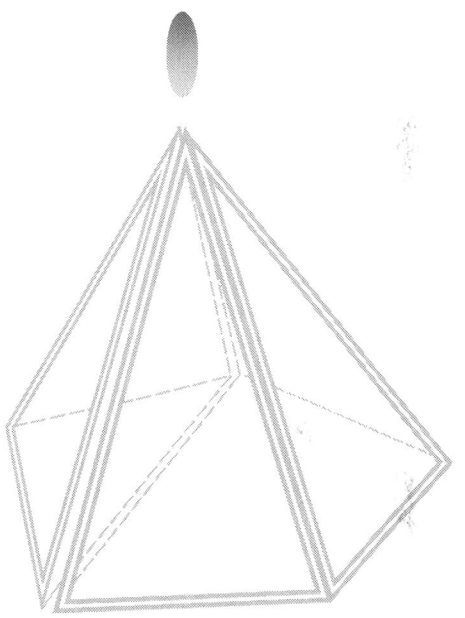

Mike Ivsin

HyperFlight Press
Boston

In Memory of Ron Reagan

Quantum Pythagoreans Copyright © 2006 by Mike Ivsin
All rights reserved
"Look inside" copy made available to search engines for phrase, keyword, and index search
For fair-use please visit Fairuse.Stanford.edu

ISBN-13: 978-1-84728-848-6
ISBN-10: 1-84728-848-0

First and second printing 2006 on paper media only

Published in Boston by HyperFlight Press
HyperFlight.com/Quantum_Pythagoreans.htm

Printed by Lulu Inc., Morrisville, NC 27560
Lulu.com/content/265736

Front cover painting by William Blake
Lessing J. Rosenwald Collection, Library of Congress.
Copyright © 2006 The William Blake Archive.
Used with permission.

Cover design by Mike Ivsin
Back cover portrait by Maru Prokop Fried, 1967
Used with permission.

Book spine HyperStates design used with permission

Editing by Theresa Welsh, Theresa@explainamation.com

Numbers, operators, and degrees of independence facilitate creation and organization of the real environment. The explanation and application of quantum mechanics on atomic and cosmic scales is suggested by the Pythagorean tradition.

1	MOMENTUM	7
2	STAR NUMBERS AND OPERATORS	25
3	THE VIRTUAL DOMAIN	109
4	GRAVITATION	199
5	THE HOUSE OF PATTER AND OCTAVIA	269
6	MOVE THE GALAXY	285
7	GENESIS	315
	APPENDIX	335
	BIBLIOGRAPHY	358

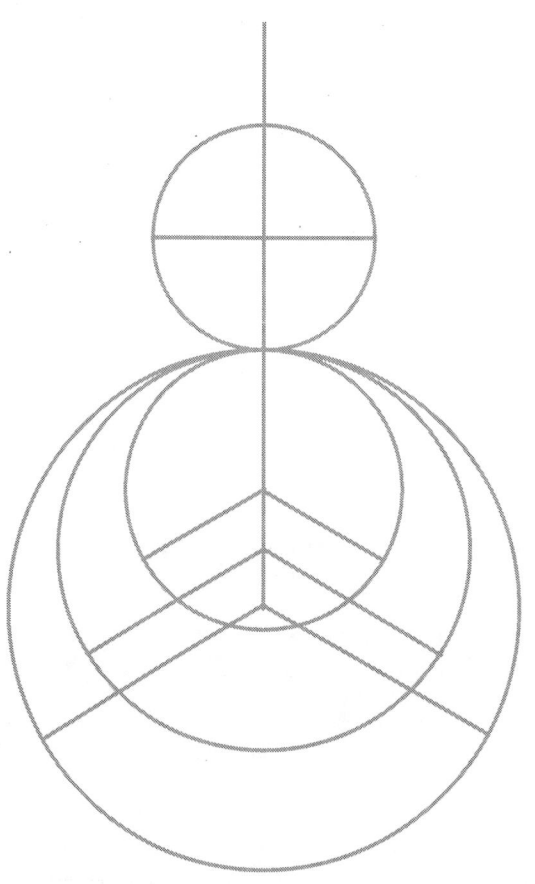

GATEWAY OF THE HEAVENLY SPHERES   8
CREATING MOMENTUM   11
ABSOLUTE AND RELATIVE MEASUREMENT   17

Momentum

     Momentum is about moving things – getting your car or your ship from here to there. Energy must be imparted on an object in some direction to get it to move. Having acquired speed and direction, the object now also has momentum. The faster it moves, the more momentum it has. Momentum is a type of energy that is conserved, for the energy that is put into speeding things up is recovered when these things slow down.

     Momentum is very special. Mathematically it is but mass multiplied by its velocity. Yet behind all momentum there is a wave, a wave that also raises the consciousness of man every few hundred years – a wave that can take you places. Twenty six hundred years ago Pythagoreans originated the idea that a moving body produces a tone – that is, a frequency – which at the time of the Renaissance grew into a cosmology of musical heavenly spheres. Wondrous it may be and the orbits, the stars and the moons are all there ready to be admired. If there are any changes coming it is, perhaps, because we could admire some of these things close up. Perhaps we could make some of these things ourselves.

● ○ ○ ○ ○ ○ ○   Momentum

Gateway of The Heavenly Spheres

Galileo Galilei explored momentum by observing a pendulum and he derived the mathematical formula that describes a pendulum swinging on a fixed pivot. Galileo rolled balls down and up an incline, investigating the conservation of momentum, by now one of the accepted physical laws. A ball rolling down from a particular level could make it back up to the same level over a gentle or a steep incline. The rolling ball, then, carries and conserves its moving energy, ready to return the energy sooner or later, gradually or suddenly. When on a level plane in the absence of friction, Galileo noted, a ball could keep on rolling forever. Any change in momentum Galileo attributed to force. While force manifests as acceleration, torque or pressure, it is not apparent how and why force arises among the nature's inanimate bodies.

Isaac Newton developed the idea of mass inertia, which allowed him to formulate the resistance a body offers when it speeds up and its momentum increases. Newton also experimented with light and, using a prism, described light as being corpuscular. In what Newton calls the crucial experiment, the prism elongates a small circle of white sunlight to reveal bands of different colors. These colors could not be divided further into other colors, yet they could be put back together into the original circle of white light. A century later, in the early 1800's the makeup of light was determined to consist of waves and light's color was recognized as light's wavelength. Another century later, the quantum theory of light gave merit to seeing light as a multitude of individual wave packets. Each wave packet of light spans some spatial distance and is called a photon. Indeed, the prism separates individual photons from a mix that is dependent on the wavelength of each and every photon.

In his work on gravitation Newton postulated the concept of absolute space and time. Newton explained the need for absolute space and time on several occasions but over the years the absolute aspect of space and time was left behind, becoming an artifact of Newtonian physics. Some two hundred years later, however, light was found to have absolute speed because light always traverses a particular distance in a particular time period. In 1887 Albert Michelson and Edward Morley sent two crests of light in separate directions at the same instance. They measured lightspeed of both crests and found them to be the same regardless of whether the crests were sent with, against, or perpendicular to the velocity of the earth's motion. Michelson and Morley reached the lightspeed constancy conclusion on the basis of their instrument's accuracy, which at the time was good enough to register the differences in earth's speed. Michelson and Morley's 1887 interferometer experiment validated the fact that the velocity of the light source has no effect on lightspeed, and light always leaves the source at a particular and constant speed. Any stationary or moving observers can now take it as proven that light entering their detectors always propagates at one value of lightspeed. A rough value for lightspeed was known in Newton's time and, while Newton discussed lightspeed without reservation, he did not reach a conclusion about light's momentum. Today there is little disagreement that photons of light have no mass. For some peculiar reason, however, photons of light are thought to carry momentum and, because of this, light ought to exert physical pressure at the surface of a mirror when photons reflect from the mirror.

We will present a new facet of light in the upcoming chapters. Although photons do not carry the punch of momentum and cannot push a mirror, photons of light can and do create motion at absorption and in consequence create momentum. Light, however, cannot create momentum at reflection because light is not absorbed at reflection. Although mainstream science claims that light produces pressure at reflection, no direct experimental measurement has been done. For over forty years now the opportunity exists to direct a laser beam at a mirror and measure light's presumed pressure directly and conclusively. On the one hand it is easy to see that light cannot move a mirror. On the other hand, mainstream science continues believing, with only their equations, that light does impart pressure at a mirror. It is reasonable to think the experimental and direct confirmation of light's pressure at a mirror would be done quickly because it would confirm century-old theories and validate much of the effort going on today. It did not happen. In reality light does not exert pressure at a mirror but pressure is still officially believed to exist even though it was never measured.

After many historical trials and tribulations we can all agree that energy is conserved and, because momentum is a moving energy, momentum is conserved as well. The perpetual motion machine cannot happen because one cannot take some amount of momentum and forever

● ○ ○ ○ ○ ○ ○ **Momentum**

produce other forms of energy without losing the machine's moving energy. The conservation of energy, however, not only allows but also guarantees anyone and any thing to forever change one form of energy into another. Energy is indestructible. Similarly, the conservation of momentum allows anyone and any thing to forever increase or decrease a body's momentum through collisions where the total momentum of all colliding bodies that makes up the total moving energy is conserved. Collision, however, is not the only way of transferring momentum. Gravitational attraction increases momentum of a body whenever a body moves faster. Yet an essential question remains. There is no problem conserving momentum of rolling balls, lumbering trains, speeding rockets and orbiting planets; but how could, and why would, that very first thing begin to move in the first place…

In 1897, ten years after the lightspeed measurements yielded constant lightspeed values, J. J. Thomson discovered and characterized the electron, which he speeded up and then let coast toward a screen. Having some mass, a moving electron has momentum and different energies accounted for different departures from the electron's path. Yet, twenty-five years later Louise de Broglie postulated that the electron could transform its momentum into vibration where momentum and vibration are two different forms of the same electron's energy.

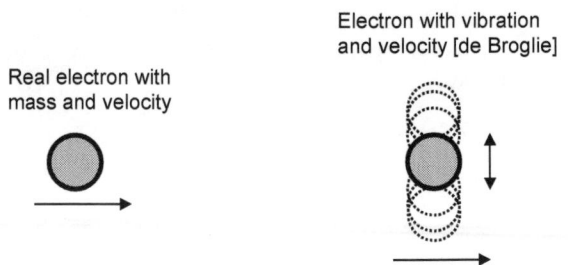

Real electron with mass and velocity

Electron with vibration and velocity [de Broglie]

Two energy equivalent states of a moving electron

Because the electron must be moving to have momentum, we can expect the moving electron to also acquire vibration. A vibrating electron could then also have wave properties. When moving at faster speeds, the electron would have a faster vibration and a shorter wavelength. Experimental confirmation that an electron has a particular wavelength came in 1927 when Clinton Davisson and Lester Germer

10

made a crystal of nickel the target of electrons. Sending electrons toward the crystal at a particular velocity validated that electrons have wavelength, and the actual value of the wavelength was in accord with known crystal lattice dimensions. It started as simple energy equivalence and grew into a discussion that continues to this day. Suddenly, a gateway opened up that some suspected was there all along – momentum of moving matter and vibration are two forms of the same energy. Best of all, the two forms of energy can also *reversibly* transform and manifest in two different ways.[1]

We will begin creating momentum by first engaging photons and electrons. By bouncing both the speedy and steady photon we will resurrect Newton's concept of absolute space and time. Riding the electron wave we will uncover the mechanism of energy conservation. These processes are not about changes for changes' sake, but they are about transformations that serve some purpose while adhering to the fundamentals of energy conservation. The idea is to open the magical forum where the creation of movement arises with the force of gravitation.

Creating Momentum

"In the beginning all but the spirit was still..."

When two bodies are at rest, their mutual as well as their total momentum is zero. Mutual momentum is the momentum bodies have with respect to each other and, since the distance between them remains the same, the mutual momentum is zero. If you were to set up an observation post on one of the bodies, you would be able to measure whether the bodies were spinning or orbiting each other. When nothing is moving there is no momentum no matter how one looks at it because momentum is a kinetic – that is a moving – energy.

---

[1] Louise de Broglie relation between electron momentum **p** and vibration frequency **f** is **p** = **f·h** where **h** is Planck constant named after Max Planck. Equal sign indicates reversibility.

● ○ ○ ○ ○ ○ ○ **Momentum**

If bodies are not moving and their momentum is zero, how could these two bodies start moving? Given that bodies in the universe are moving without physical collisions, can these two bodies commence moving without something colliding with them? There is a bit more to this than a cosmic analogy. Should one body collide with a moving body, it leaves the question as to how *that* moving body got moving before it was involved in a collision that caused something else to move. If we are to explain the creation of movement, we cannot presume that something is moving already. If momentum is to be conserved and it is zero, then momentum must also be zero later on. Accordingly, if the total momentum is zero to begin with and also to end with, is it possible to get moving without a collision?

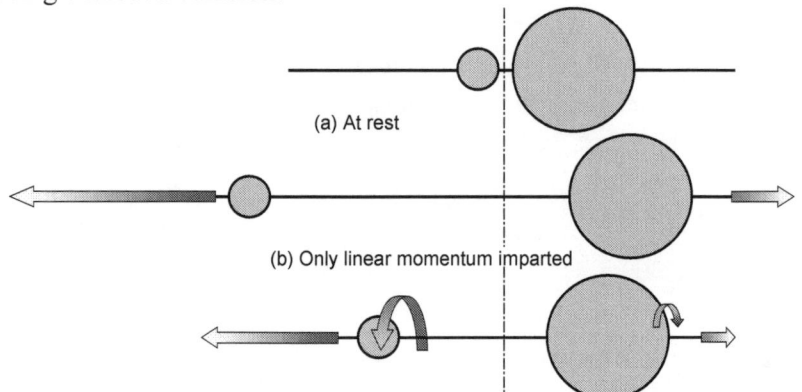

Two-body system at rest (a), and with identical amounts of added energy (b) and (c)

Giving each body the same amount of energy that is in the opposite direction creates movement while conserving the total momentum. Both bodies considered together continue to have no total momentum because each body has just enough energy to bring the other body to rest where the resting state is the initial state. Summing the momentum of both bodies yields total momentum, and the total momentum remains at zero because the momentum of the first body is the same and in opposite direction to the second body. While the bodies are moving, the imparted energy is in the form of moving energy and the imparted energy is conserved. The moving energy of both bodies is equal to the total imparted energy. Some or all of the imparted energy can be recovered at any time by slowing one or both bodies down or by

12

bringing the bodies to rest. By dividing the application of energy in half, bodies speed up in the opposite direction while the energy as well as the total momentum are conserved at all times. Bodies commenced moving while conserving momentum because the total momentum before and after the application of energy is the same even though the total momentum happens to be zero.

There is now a system of two bodies and with it a condition. Both bodies can have any amount of mutual momentum as long as the total momentum in this two-body system remains the same. This condition can be stated as a postulate: Energy can be applied to move bodies if the total momentum is the same after the application of energy as before the application. To uphold the postulate, at least two bodies are needed. To state the postulate in the most general terms is to say that no energy can ever be converted to moving energy unless the total momentum is conserved. If you think ahead about other possibilities, you can also uphold the postulate if you spin a second body off of one body and you can think how you would create dual suns from a single sun, for example. Also recall Andre Ampere's discovery that gas molecules consist of two or more atoms. Such plurality will be put to good use later on.

Galileo first wrote about experiments conserving momentum of bodies that are already in motion while Newton expanded on it by postulating the necessity of equal and opposite forces in the creation of movement. The conservation of momentum among real moving things has been experimentally confirmed many times and so, with a good foundation, we are going to move on to the creation and the distribution of momentum. It is technically easy to arrive at a conclusion that energy must be split in half before it can be applied to two bodies in the opposite direction. Now we are on a lookout for entities that can do that.

● ○ ○ ○ ○ ○ ○ **Momentum**

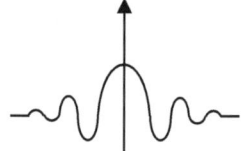

Photons symmetrical energy distribution

As it happens, photons are always symmetrical about the vertical axis – that is, photon's energy is always evenly distributed in a fifty-fifty fashion about the vertical axis.[2]

One unique property of light is that light's propagation always maintains symmetry about an axis. Direction of light's propagation can change instantly and the even – that is, symmetrical – distribution of light's energy remains instantly and evenly distributed as well. To test this, take a point source of laser light and direct it through any and various optical arrangements including gratings. An image having axial symmetry always results.

Momentum can only be, and has always been, created when two bodies move or rotate in the opposite direction. The recoil of a rifle and a motor's outer housing rotating in the opposite direction of the rotor are but two examples of the conservation of momentum during movement creation. Light's energy is evenly distributed about an axis and this points to a mechanism that allows light's absorption to be accompanied by the creation of movement. Nuclear radiation causes damage to living organisms but only recently is the chain of radiation damage understood. A high energy X-ray photon breaks up the molecule of water when it is absorbed in between hydrogen's proton and electron, as illustrated below.

---

[2] Carl Jung compiled a number of symbols he called archetypes, a circle and a wheel being prime examples. Jung's observation was that archetypes are so fundamental to the psyche of a person they transcend cultural differences between peoples. A photon's energy distribution shape with its axis can be added to the list of archetypes, for this shape appears prolifically as symmetrically positioned or "kissing" animals such as snakes and symbols with dual heads or symmetrical wings.

Quantum Pythagoreans

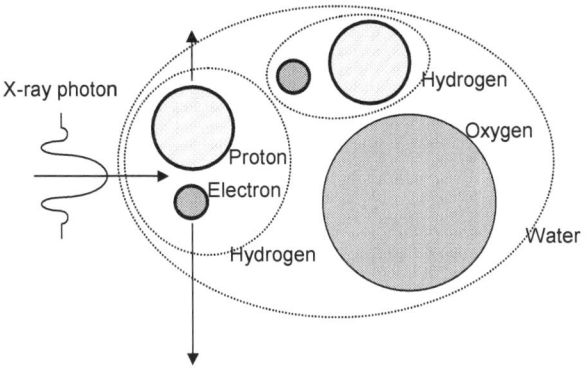

X-ray radiation breaks up a water molecule and creates a free radical

A proton and an electron are forced apart from each other in the framework of momentum conservation. They each move in the opposite direction thus leaving OH, the free radical, to cause damage.[3]

The creation of momentum is also about the conservation of energy because any form of energy must be conserved, even though it is transformed from electrical to moving energy, for example. Mathematically, the distribution of momentum in the framework of momentum conservation is a necessary but not a sufficient condition of momentum creation. The distribution of momentum is a prerequisite to momentum creation because for every new way of distributing momentum there is a new opportunity for creating momentum.

A body is spinning when it rotates or orbits another body. Angular movement is about angular momentum because spinning is about moving energy as well. There are, then, two kinds of momentum: linear and angular. Linear momentum has symmetry about a straight line – the actual physical or mathematical entity constantly repeats as it translates. Angular momentum has symmetry about a point in that the physical or mathematical entity repeats about a pivot in a circle.

---

[3] OH is a free radical called hydroxyl. OH does not have a net charge but its compulsive reactivity stems from the unpaired electron that oxygen is now looking to pair up. OH is the source of damage to cells of living organisms because the hydroxyl free radical rapidly damages a neighboring cell, which also starts a chain reaction of new damage. A more thorough description is in Nick Lane's *Oxygen*. Hydroxyl is one of three free radicals between oxygen and water.

● ○ ○ ○ ○ ○ ○   **Momentum**

An object moving at constant velocity has linear motion also called translation, which is solvable with linear variables – that is, variables of distance in simplest problems are the power of one. Angular motion requires quadratic equations for a solution and in the simplest case the variables of distance are the power of two. An object undergoing gravitational acceleration in a straight line also has a quadratic relationship to the variable of distance. A compass is a geometric tool that facilitates rotation about a point while an unmarked straightedge provides translation. A compass also doubles or halves any straight distance or any angle. All linear and quadratic equations are solvable with a straightedge and a compass.

In an unusual and possibly fortuitous twist, the number π does not come out as a solution from linear or quadratic or any higher order equation. Using a straightedge and a compass, π can only be constructed approximately. Number π is routinely defined as a ratio of the circle's circumference to its diameter, but it is not possible to put the value of the circle's circumference into the computer to get π. Such definition of π is then not very practical because both the circumference of a circle and π are numbers with an infinitely long mantissa – and only by summing an infinite sequence of ever decreasing terms can we arrive at π. Another way of looking at π is that at some starting point it is possible to traverse either a straight distance **2** or a semicircle distance π to get to the same spot at the end of distance **2**, which now also becomes the diameter of the circular distance. Archimedes discovered that the circumference and the area of any circle, as well as the volume of a sphere and a cone, relate to its diameter through the same number π and that is how π became a constant.

When energy is imparted on bodies to change their speed or direction, one can produce any combination of linear and angular momentum but both the linear and the angular momentum have to each add up to the same values before and after the impart of energy. When two electrons spin in opposition they store the extra spinning energy locally and without net angular momentum. If two galaxies were to have a common axis of rotation, and many do, a good guess is that they spin in opposition.

Qualitatively, momentum is linear and angular. Quantitatively, total momentum is always zero while mutual momentum can have any value. But of course, the total momentum in the entire universe is zero. The universe can be expanding but the universe as a whole is not translating nor is the universe rotating.

Absolute And Relative Measurement

When observers are measuring a variable and obtain the same value, the measurement is said to be absolute. Temperature and pressure measurements are absolute because all observers' results are in agreement. A particular parameter's value may be absolute but the word 'absolute' is used in the framework of measurement only. Absolute measurement means there is no disagreement as to the parameter's present value but it does not mean that such parameter exists forever and could not be transformed.

At first glance the measurement of some object's velocity is not absolute because slow and fast moving observers obtain different velocity measurements. The results of velocity measurements are not always in agreement and then the values are relative to the person making the measurement. Momentum measurement is affected the same way because momentum is velocity of an object leveraged by object's mass. To make a relative velocity measurement into absolute measurement calls for a reference point that is placed at the point of absolute rest and then all velocity measurements with respect to the reference point will become absolute. However, the measurement of lightspeed results in a value that is always the same even if it is made by slow or fast moving observers. Lightspeed measurement is also the same if the source of light moves slow or fast. The source of light can be going at any speed and at any direction but once light leaves its source it continues propagating at lightspeed in the direction the source is pointing at the instance of light's release. The results of the 1887 Michelson and Morley lightspeed measurements established that the velocity of the light source is not added to or subtracted from the velocity of light, and light propagates with constant speed once it leaves its source. The velocity of light is not affected by the velocity of the light source pointing with or

17

● ○ ○ ○ ○ ○ ○ **Momentum**

against or on a diagonal or in any direction. For the purpose of measuring lightspeed, the reference point is not needed because lightspeed is inherently absolute and lightspeed is the absolute reference for the parameter of speed.

The existence of absolute spatial distance and absolute time is obtained from the absolute speed of light because light in the vacuum of space always propagates at a particular and constant speed. Having the absolute speed of light, an instrument can be built such that absolute distance and time can be extracted from the absolute speed of light. The construction of an absolute clock calls for a light source with detector, and a mirror placed at a fixed distance that reflects light back to the detector.

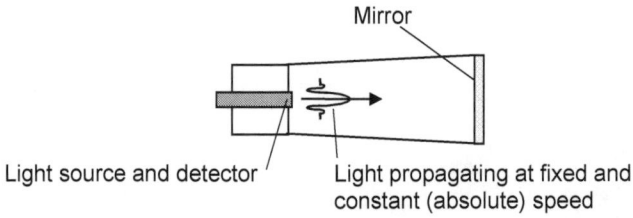

Light source and detector   Light propagating at fixed and constant (absolute) speed

Absolute clock with light source and mirror at fixed distance

The detector's role is similar to that of the mirror, for the detector also sends light back out to the mirror for repeated bouncing and continued counting of the clock's ticks. This clock is absolute because fast or slow speed of the clock device as a whole will not change the time of light's round trip. If the clock device direction of travel is in parallel to the light path then the duration of the round trip of light will be the same regardless of the speed of the clock device. Even though the clock device is moving at some velocity, such velocity is not added to the speed of light and the light always leaves the light source at one constant speed. The clock device moving in the direction of light cannot speed up light. The clock device moving opposite of the direction of light cannot slow down light. If the clock device is moving in the same direction as the direction of light, the time it takes to reach the mirror is somewhat longer but the return trip is shorter by the same amount.

Regardless of the speed of the clock device, and be it to the right or left or fast or slow, light will always complete the round trip in one fixed time duration. If the clock's direction of travel is perpendicular to the light path then the round trip will also be the same because the velocity of the clock device is not added to the speed of light.

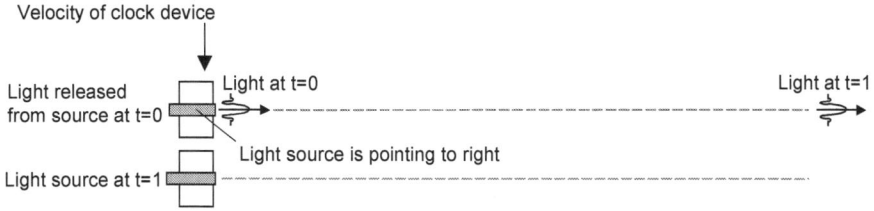

Light source downward velocity is not added to the velocity of light

If light crest round trip were not the same going with and perpendicular to earth's velocity, the Michelson and Morley interferometer would have measured it. See the Appendix. The orientation and velocity of the clock device has no effect on the timing of the clock. Similarly, if the clock device is moving on the diagonal, no component of the device's speed will be added to light speed; light will propagate in the direction the light source is pointing at the time of light's release. Moreover, if light's path is perpendicular to the direction of movement of the clock device then such movement can be accelerating or decelerating while the clock remains the absolute clock.[4]

Absolute time is thus *derived* from the constant and absolute speed of light. You now send two observers, each with an absolute clock and a light source on separate ships to different parts of the universe. These observers send light signals to each other at will, and they will always agree on the distance light traversed when they receive the signal because they both have the send and receive times in absolute terms.

A reference point having absolute rest may seem ideal but if such a point is in one spot in the universe it would be most difficult to extend

---

[4] If the mirror is a large distance away and the speed of the clock device is perpendicular to the clock's light path then the returning light could miss the detector. This then imposes limits on the rate of acceleration and also on the geometries of structures relying on an absolute clock for their existence. Only the direction that is perpendicular to the clock's light path can be used for acceleration to guarantee the constancy of light's round trip. The clock device or a structure relying on an absolute clock needs to rotate and tilt such that the direction of acceleration is always perpendicular to light's path.

● ○ ○ ○ ○ ○ ○   **Momentum**

the reference to another spot in the universe. Constant lightspeed, however, extends such reference to any point in the universe. Technically, the entire universe is marked with a grid of known distances with arbitrary resolution while each and any point also has local time that is also the absolute time. An object can move between two points with speed that is conceptually slower or faster than the speed of light because the parameter of lightspeed merely established repeatable – that is absolute – distances. Spatial distance markings do not modify the properties of space and so the spatial distance and time can be visualized as overlays onto space. The constancy of lightspeed results in knowledge that absolute distance and absolute time can be constructed. Lightspeed, then, is the one and only building component in the creation of a *formal* system that is the real system. Any event happening at any time has absolute spatial distance and time coordinates. Moving or stationary objects can be differentiated. Statements such as 'sometimes later' or 'event A before B' hold because the grid system of absolute distances and absolute time allows observers to agree on precedence. Another way of saying it is that each and every object has an unambiguous history of its movement. The absence of ambiguity defines the formal system. Absolute spatial distance also guarantees absolute gravitational force because there exists but one distance and, therefore, there exists but one gravitational force between two bodies spanning such distance.

Absolute spatial distance and absolute time do not dictate a continuum of spatial distance and time, which is an artificial and limiting, rather than simplifying, construct. An object disappearing at some spot in the universe and reappearing at another spot in the universe does not violate the absolute coordinate system of absolute spatial distance and absolute time but violates the presumption of the space-time continuum. A continuum forms among two or more variables that cannot be separated. It is apparent that continuums can form only if energy is included. Momentum is a continuum of real mass and velocity because momentum is also a moving energy. The conservation of energy then guarantees that a continuum remains a continuum. A continuum cannot be destroyed but can be transformed in the framework of energy conservation. Spatial distance and time individually or in combination do not contain energy and in consequence do not form a continuum. The Planck constant, for example, does contain energy.[5]

Spatial distance is often called just space. Spatial distance is about separation between things rather than about the in-between or the content of the in-between. Prior to quantum mechanical discoveries at the beginning of the 20th century, space always referred to spatial distance. With the advance of quantum mechanics, space between things acquired new properties and more expressive names such as quantum vacuum were introduced to further differentiate space from spatial distance. Space and spatial distance need to be differentiated. Spatial distance is a formal construct introduced by lightspeed as an overlay onto the quantum vacuum; it is measured in feet or light-years or some such unit of distance.

An argument can be made that relative motion is a fully functioning subset of absolute motion. A relativity postulate can be made that even though one must always establish a frame of reference for relative motion to work, relative motion simplifies things in some way and the absolute reference is then not needed even if it is available. Consider, however, that every accelerating electron radiates electromagnetic energy and is the reason that all wireless applications work. Should relative motion be sufficient, an accelerating frame of reference passing by the electron would make the electron radiate and elicit energy from the electron – yet this is not possible. Multiple references accelerating past the electron at different accelerating rates would then elicit different energies from the same electron at the same time. None of this is possible. In the quantum mechanical environment of electrons and photons, absolute measurement is required and is available as well.

Relative motion is a special case of motion and cannot be generalized. Relative motion can only be used with electrically neutral objects having at all times well-defined mass and position. Relative

---

[5] Planck constant is the product of energy and time. There will be more to say on time and transformations in the upcoming chapters. For example, time remains absolute only as long as the variable from which time is derived remains absolute.

21

● ○ ○ ○ ○ ○ ○   **Momentum**

motion framework is thus restricted to collisions among objects. Relativistic framework is sufficient during collision because energy is summed up and the total divvied up among objects. In general, however, when an object speeds up as a result of a particular imparted amount of energy, the de Broglie wavelength of the object changes because the wavelength is based on the object's new moving energy. Because the relativistic postulate allows the frame of reference to be placed with any object then such objects would carry momentum relative to the frame of reference: Objects now have de Broglie wavelengths that were recalculated with respect to the new frame of reference but such wavelengths no longer reflect the actual energy imparted to the objects. When discussing the dual slit experiment in the upcoming chapters the wavelength of each individual object becomes important. The individual object can carry positive or negative charge or be neutral, but knowing the wavelength that is commensurate with the actual moving energy is necessary if the wave interference phenomenon is to be predicted through computational means. The actual and absolute moving energy is expressed through de Broglie wavelength whether the object's speed is constant or changing. The wavelength is representative of the actual energy imparted on the object – and if not changed and corrupted under relativistic presumption – the mathematical use of the correct wavelength is in agreement with experimental results.

    Electrons and photons of light move and interact in our everyday space, and in simple experiments the absolute spatial distance and time parameters do not introduce much complexity. For example, in the dual slit experiment that uses electrons or other particles, the frame of reference is always placed at the dual slit because its placement with the moving particle would not work. Only by placing the frame of reference at the dual slit will the math be in agreement with experimental results.

    With the application of lightspeed, absolute measurement becomes available; the absolute spatial distance and absolute time can be constructed, and the absolute speed measurement of any object obtained. Formal systems as well as the quantum mechanical system both require absolute spatial distance and time. In the future chapter on the electron, additional electron parameters will make the electron complete. Absolute rest will come up again because momentum conservation would be simplified for discontinuous movement through space.

Newton postulated absolute spatial distance and time as existing. In our baseline, we established both the absolute spatial distance and absolute time as derived from the existing absolute speed of light. The difference may become a mere point of a philosophical discussion because the absolute spatial distance and time are *constructible* and, because they are available, they exist. It will turn out that the gravitational timing, which is inherent in Newton's gravitational constant, is absolute but derived from a mechanism other than lightspeed. Newton's reach extends to this day. In time, the absolute distance and time will become easy to apply and Newton's call for absolute distance and time will become natural once again.

NEWTON'S STANDING WAVES	26
INFIDELITY OF THE INTRACTABLE	32
GEOMETRIC AND ARITHMETIC COMPUTING	34
Symmetries and Dimensions	*35*
Geometry and Arithmetic	*42*
Tractability Revisited	*44*
RATIOS PLAY MUSICAL ORBITS	46
EXTRA MOON	57
IRRATIONAL AND DIVINE	59
NUMBERS EXPRESS VARIABLES	74
Dependence And Independence of Variables	*74*
Causality And Priority	*83*
AT THE GATE TO THE CAVE OF THE SHADOWS	86
Dual Slit	*87*
Branching and Spreading	*103*

Star Numbers and Operators

Planetary orbits are in a plane and planets take their time before they repeat. Repetition is a period every wave needs. But it also happened that planetary orbit times relate to each other in ratios that make multi pointed stars. Moreover, orbit ratios are musical ratios and planets could then be making music over time. What makes a planetary ratio harmonious is the geometry of the circle and the Pythagorean gem, the number that perfectly fits in the circle.

In the atomic environment the waves wrap around and make a closure around the nucleus. If the number two divides a circle exactly, two full waves can wrap around the core without a remainder and the closure of such wave would be stable. Some numbers, in and of themselves, could be the building components of atomic orbitals.

Operators work with numbers. Forces move things and forces are operators. Numbers and operators need each other.

Newton's Standing Waves

> *"Hypothesis non fingo."*
> Newton
>
> *"I frame no hypothesis."*
> Newton translated
>
> *"I do not make hypothesis out of thin air."*
> Newton interpreted
>
> *"I don't spin it to create hypothesis."*
> Newton restated

The colors of the rainbow and the colors produced by a prism are the same and this was not lost on Newton. It is said his book *Optics* was more popular than his *Principia*. Newton observed that the image produced by all lens-based refracting telescopes suffered from color separation and realized he was dealing with the intrinsic property of a prismatic shape of a lens rather than with poor lens quality. Believing that color separation could not be easily rectified, Newton designed a telescope with a large curved mirror instead of lenses. His solution was to reflect the image to come out of the side of the telescope body and thus separate the incoming and the amplified image. It was this homemade telescope that earned him the election to the Royal Society.

There is something remarkable about Newton's staying power over the centuries. Even during the light-is-really-wave debates, Newton was able to stay above it because the wave nature of light did not invalidate his experiments. Newton did not endow light with mass. Newton originally described mass (materia) as "proportional to the density and the bulk co-jointly," which echoed the existing consensus. Newton later formalized the mass of a body as proportional to the inertia, or resistance, a body offers to being set in motion by a given force. Newton generalized mass when he enhanced the static definition of mass – mass weight – by introducing the dynamic characteristic of mass – mass inertia. Yet, Newton did not make a case for light having mass or, in his term, as being material. Newton's conclusion on light was not just luck or a guess and his choice of the word *corpuscular* for the characterization of light then stands for a form of embodiment or packet grouping that conveys a *cohesive* attribute of light. His choice of a word that is different from mass, coupled with his knowledge of mass, clearly

Quantum Pythagoreans

indicates that Newton thought of light as a unique entity. In a logically consistent Newton update, light consists of mutually independent and indivisible flying packets, or photons, of particular colors.[6] One of Newton's lasting legacies is experimental confirmation.

When a curved lens is placed on top of a flat glass surface, a series of concentric and colorful rings appear that are called fringes. Newton set out to explain this phenomenon while making the best of his English. What he called *fits* when explaining the rings of optical fringes are today called *nodes*. But Newton's fits are possibly more descriptive than nodes because fringes change color when the half-multiple of a wavelength *fits* into a gap. Newton's power of conceptualization also becomes apparent as he is describing a phenomenon without the established model of a standing wave.[7]

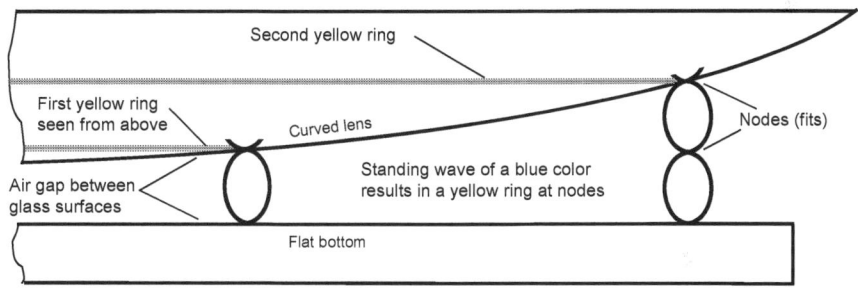

"Newton's ark" allows light's wavelength measurement without the parameter of lightspeed
Newton rings in profile

Newton measured the distance from the first ring to the center where the lens touches the glass on the bottom. He knew the diameter of the grinding tool that produced the lens and calculated the height of the air gap, which for the yellow ring he reports at 1/80047 of an inch. Newton certainly is the first person to measure the wavelength of light.

---

[6] In Newton's hand, "...Light is it self a heterogeneous mixture of individually refrangible rays."

[7] Standing wave illustration is similar to the Egyptian glyph signifying 'closure.' Newton was a student of ancient Egypt and Egyptian cosmology in particular. This is not to say that Newton would recognize the 'closure' glyph as a standing wave even if he had seen it. 'Fit' and 'closure,' however, do have common traits.

● ● ○ ○ ○ ○ ○   **Star Numbers and Operators**

To top it off, when Newton reports on the measurements he most unexpectedly and correctly doubles the air gap to encompass the entire wavelength. Because the value at nodes is zero for a particular color, the location of the node denotes the absence of such color. A yellow color results from the absence of a blue color, for yellow consists of but red and green. By calculating twice the air gap at the first yellow ring Newton actually measured the wavelength of the blue color. Newton's measurement falls into the lower range of the violet color, which is about 25% off the modern blue color classification.

During the early twentieth century the electron also became a wave, and the model of the first quantum mechanical atom was then possible. The fundamental assertion is that the waves of electrons form quantized orbitals around the core. Using Newton's easy terminology, the orbital forms at the fit. Orbitals are quantized because orbitals form only when the wavelength fits the distance around the core. Newton's fits make sense as both the noun and the verb. Newton's fits are usable with linear and angular geometry – as a standing wave and an orbital.

Relative motion is a notion that did not sit well with Newton. Relative motion declares that one thing passing another is equivalent to the other thing passing the first and no one is wiser as to what is moving and what is not. Newton's logic regarding absolute spatial distance and time was simple enough: If you are spinning on a merry-go-round you know you are spinning and when you stop you know you are not spinning. Relative motion is not supposed to give any clues as to whether it is you or the earth that is spinning. Anybody ignoring geometries or acceleration can make his or her theories but continuing generalization will eventually run up and against the funky merry-go-round. Similarly, if a person is accelerating along a straight line and has a bucket full of water on board, you know the person is accelerating when the bucket overflows. If the bucket does not overflow you know it is the reference that is accelerating going past the person with the bucket. Although Newton knew that the creation of absolute reference would be good for the whole universe, he admitted he did not know *where* to place the reference. Because of lightspeed constancy, the reference is everywhere and anywhere.

Absolute rest would do just as well for staking the reference, but absolute rest was and remained in the realm of conjecture. Unable to express absolute and relative velocities quantitatively, Newton in his *Principia* nonetheless differentiated between absolute and relative velocities qualitatively. The possibility that there could be a speed measurement that would actually result in the absolute value did not exist even as a speculation and not until the 1887 Michelson and Morley experiment was the possibility of the absolute speed measurement even contemplated. Lightspeed constancy now allows the construction of the computational framework for absolute spatial distance and time.

Newton wanted to explain gravitation and he was convinced that gravitational forces between bodies act instantaneously on each other.[8] Newton's formulation of the conservation of momentum states that if one body acts by force on another body, the other body reacts with the same and opposite force on the first body. The action is accompanied by an instant reaction, which means that forces arise concurrently while also increasing or decreasing concurrently. When forces arise concurrently then Newton is also saying that the conservation of momentum holds at every conceivable instance of time. Instant action is accompanied by equal and opposite instant reaction and any net momentum such forces may produce will add up to zero. Instant action is easy to validate for physical collisions and Newton went to great length in his *Principia* to show that the concurrent forces of action and reaction happen across distance as well. He used rotating balls linked by a string and computed forces at the center of rotation as being the same and in the opposite direction.

Instant action is characteristic of light's change in velocity during reflection and refraction, and instant action is also prominent in quantum mechanics. Newton struggled with ether, on again, off again. Newton knew something must facilitate gravitation – in his words there must be an agent – but then again ether would be in the way, slowing

---

[8] Johannes Kepler was first to postulate the mutual nature of the gravitational force; one body attracts another the same way the other body attracts the first. Kepler's initial validation of mutual force came by linking ocean tides to the moon's periodic orbits. Rising ocean water may seem one-sided but it was the first proof of the force-acting-at-distance and the symmetry made sense because experiments with magnets confirmed the mutual nature of magnets' attraction.

bodies down and literally grinding them to a halt. How subtle should a subtle medium be? In a letter, and perhaps in the ultimate expression of frustration, Newton the expert left it up to the readers of his book to decide whether the agent of gravitation is "material or immaterial." Perhaps it was Newton's ultimate expression of his faith in humanity, for he left the problem of the gravitational agent to his fellow humans. Judging from the context of the letter, however, Newton was not happy with men of cloth interpreting scientific results.

Newton researched historical thoughts on gravitation and gave Pythagoras credit for discovering his own square law. While Pythagoras did not use arithmetic, Pythagoras knew that to double the string's vibrating frequency the force tightening the string had to be made with "square numbers." In the case of frequency doubling the force tightening the string had to be multiplied by four. Newton liked occult and esoteric writing, and appreciated that equations are not the only way of encoding and conveying knowledge.

Mathematical formalism updated and advanced Western knowledge concerning gravitation. The most beneficial result of mathematical language is its ability to lift the fog of obscuration and focus on leveraging the proven result even if the underlying mechanics still had to be taken as a given. Galileo appreciated force as something that gets things moving and he made many advances experimenting with moving bodies of projectiles and a pendulum. Galileo encountered so many different explanations of the mechanics of force he reached the conclusion that the search for the cause of force was a distraction – and he was satisfied enough to mathematically describe objects' movement once they got going. Through his measurements, Galileo realized that gravitational force results in continuous and uniform change in velocity and thus the force of gravitation produces acceleration that is a change in distance per second every second. But it was Newton once more who was able to formulate and generalize such continuing change with his invention of the mathematical differential, a component of calculus.

Newton first defined inertia as force, *Vis inertiae* in Latin, and from then on inertia is viewed as a force that resists the change in velocity of a mass object. That is one way of looking at inertia but in addition, and using Newton's term, inertia will be presented in the upcoming chapters as the agent of the conservation of energy. An enhanced description of inertia will take us closer to the mechanism of transformations.

Newton was a prolific writer on gravitation, light, mathematics, alchemy, astrology and religion. By the time he published *Optics* his telescope was more than three decades old and Newton was meticulous in only publishing conclusions that could be experimentally backed up then and there. For discussion and exercises Newton included numbered queries at the end of the book.

Newton had a powerful personality. An excellent organizer, always putting the merits up front, Newton was appointed the Warden and later the Master of the Mint, where he single-handedly restored credibility to the English coinage. In addition to impressing the edge of coins with markings to deter trimming, he also guaranteed coins' precious metal content through statistical assaying. By taking speculation out of the value of coins he increased the velocity of money and its international acceptance. He pursued counterfeiters mercilessly and took at least one counterfeiter permanently out of circulation.

Newton did not always succeed. The problem of longitude got royal attention in 1676 when an observatory was established for the single purpose of determining the longitude at sea. Better than one degree of accuracy was needed to significantly improve the navigation of merchant shipping. The appointed astronomer, John Flamsteed, had a small budget from which he was paid and from which he acquired instruments as well. Newton wanted to combine lunar observations with his gravitation equations but could not get accurate observational data quickly. Following the 1707 shipwrecking episode of four Royal Navy ships, the Parliament in 1714 offered the "longitude prize" of 20,000 pounds – yet, the offer, twelve million dollar by today's standards, did not produce tangible results for two decades. Newton was able to change the observatory's charter and actually forced Flamsteed to publish the lunar data – only to have the astronomer cry foul and later withdrawing the publication by claiming the data to be erroneous. In 1753, twenty-five years after Newton's death, a German mathematician Johann Mayer was the first to produce lunar tables with desired longitude accuracy. Even though it was watchmaker John Harrison who prevailed with his time-based method and in the end collected the prize money, the expensive and fragile chronographs would not become the ship's sole longitude instrument for another hundred years.

Lunar tables cannot be reused from one year to the next. The ongoing publishing of lunar tables showed the Zero meridian at the Greenwitch observatory, which then also established London as the world's time reference. In retrospect, Newton treated the original observatory as a Civil Service institution with Flamsteed as its only employee. Flamsteed, however, saw himself as tenured Astronomer Royal. It was the moon that validated Newton's gravitational equations while a truly practical extension of his discovery had personally escaped him. Newton never married.

Newton was first to make the reflecting telescope, which also happened to be a telescope of his own design. The Newtonian telescope is well suited to scaling and many models, including the largest, follow Newton's original design. With a possible exception of a cannon, there is no other instrument whose design continues to be built to this day, for over three hundred years after its invention.

Infidelity of the Intractable

Gravitational interactions between two bodies of any size going at any velocity are tractable. Orbits have mathematical solutions that result in a periodic path of a circle or ellipse. Flybys are tractable with hyperbolic solutions. Bodies in sub-orbital paths are tractable with parabolic solutions. Two bodies are also tractable if they are moving along a straight line and attract each other. Since there are no restrictions on two bodies being tractable, two-body configurations always have a solution. An Earth-orbiting satellite and a dual sun system are two-body configurations.

Appearance of a third body introduces chaos into a previously tractable two-body system and the mathematical solution is lost. Three or more gravitationally interacting bodies travel on a non-repeating path and any mathematical description of the motion of the bodies becomes intractable. This means that even if all bodies' position, velocity, and mass are known exactly at some point, the mathematical modeling of three moving bodies begins to deviate from the actual and in time becomes irrelevant. Improvement in mathematical methods delays irrelevancy but the continued emphasis on improvement in accuracy leads to intractability because the error in computed position becomes zero when the computing time reaches infinity. In the short run, three-body problems can be mathematically emulated only approximately and the emulation always becomes irrelevant in the long run. Another way of stating this is that gravitationally interacting bodies conserve their momentum first, while the path of any of the three bodies becomes secondary and a body can take a path with non-repeating sharp turns or loops. The inability to emulate three-body gravitational interactions is difficult to accept because the underlying and lingering question then becomes, 'How are these bodies interacting and how come the most powerful computer cannot deal with that?'

If a problem has a solution then the problem is computable – but if the solution is not available in real-time then such a problem is intractable. The solution can be proven to exist but the lack of real-time fidelity reveals serious deficiencies in methods that attempt to emulate given problems. While the lack of real-time emulation clearly indicates that the computer cannot deal with the problem, intractability is oftentimes overlooked or ignored. Overcoming intractability, moreover, presents a great gateway to the understanding of physics and our environment. Presently, the computing machine handles some problems tractably and others intractably. It is not always apparent that computations of intractable problems have to be scaled back and that the answer is always somewhere between approximate and irrelevant. A computing machine may be labeled 'universal,' but would anybody buy a "universal" machine if it takes forever to complete some jobs? A solution is really not a solution if it takes forever to get the answer.

Much theoretical work can be done, but it is apparent that the structures and the constructions of nature always apply tractable methods. Tractability, then, is the desired framework for understanding of the issues in our environment. While the three gravitationally interacting bodies relate and thus compute among themselves in some fashion, the present day computing resources cannot emulate such three-body relationships in real-time. Because even a dramatic speedup in computing power cannot resolve the inability to emulate a three-body system, herein spring the necessity, the privilege and the opportunity to

33

say that gravitational relationships are about an entirely new class of computing.

Intractability appears trivial if it is thought of as insurmountable. It is easy enough to calculate that to break a secret code head on, the code breaker needs to do a very large number of computations and a machine going at breakneck speed will take, for example, ten million years to arrive at the answer. A new machine can be developed at some cost that can be a million times faster – and this feat was accomplished in about thirty years between the 1950s and the 1980s. Hundreds of machines working in parallel could then manage to break the code in a reasonable length of time. The code-making algorithm, however, can be changed in a matter of days to make the code-breaking just as difficult as before. For some people it is a foregone conclusion that there is no way of breaking a code using brute force computing methods where every possible combination is examined for a potential solution. Nevertheless, there is a way and there are some who believe that code-breaking can be accomplished in a reasonable length of time without relying on exhaustive methods.

One of the challenges, therefore, is the answer to the following question: If it takes ten million years to arrive at another solar system, is such length of time the result of calculations of an intractable problem? If your ship were to traverse every inch of distance between here and there, does this amount to applying exhaustive methods?

Geometric and Arithmetic Computing

Solutions that apply geometric methods are obtained with compass and straightedge. Geometric methods may turn to advantage once it becomes apparent that geometry allows exact division of a circle while the arithmetic of the computer cannot even store a single irrational number such as $\sqrt{2}$. $\sqrt{2}$ is easy and quick to construct geometrically and no rounding or truncating is necessary. Some computations are tractable in geometry while the very same computations are intractable in arithmetic.

Symmetries and Dimensions

A photon's energy distribution is always symmetrical about a line that is the axis of symmetry. A line of symmetry reflects evenly to the left what it is on the right or vice versa. Photons and some other atomic components are called *even functions* for just that reason. Generically, all even functions are forces and behave inclusively – that is, forces superpose and add or subtract from each other if they are the same kind of force. Forces move things and can be called operators. Force can also be equated with Aristotelian *potentia* – force can manifest as movement but force needs something to operate on before movement can happen. Oftentimes, forces are called fields but forces form fields only when forces are shaped with geometric constructs and there exist no fields without them. A field, then, is a force that, because of geometric constructs, results in a computable 0, 1, 2, or 3 dimensional description of force.

Zero-dimensional symmetry is symmetry about a single point, which may carry other advantages – and yes, there are atomic entities having symmetry about a point called *odd functions*. Electrons and protons are in this group. Generically, members of the odd function group are real things and, once localized, behave exclusively with each other – that is, real things displace each other instead of passing through each other. Localized real things are particles. Forces put particles in motion. Fields are structured forces that put particles in computable motion, which for some geometries also results in tractable motion. Due to the exclusive nature of particles, they can be stacked together to form stable geometric constructs.

The odd function operates under mathematical rules that result in the impossibility of a solution if two odd functions were to occupy identical space. The collision, then, has an underlying mathematical form. Even function, on the other hand, allows mathematical result where two or more even functions overlap and superpose with each other. In general, a mathematical function can be even or odd, or neither even nor odd. The atomic framework, however, has only the even or the odd function. Most mathematicians are well aware that for one reason or another the physical world is but a subset of their mathematical tools –

all the while knowing that some obscure mathematical relation will sooner or later find an application and become famous.

Two-dimensional symmetry is symmetry about a plane, which requires an addition of another – the third – dimension. Looking into a mirror explores the two-dimensional symmetry that reverses left and right, but the two-dimensional symmetry keeps the up and down invariant. In front of the mirror, both the left-hander and the right-hander can see each other's thumbs point in the same direction and that becomes the *up*. The left-hander is talking about the left-handed screw while the right-hander is talking about the right-handed screw but both will get the job done. Both thumbs point in the same direction in front and, also, on the image side of the mirror where the right hand becomes the left hand and vice versa. The conclusion is that the *up* (or *down*) is invariant under two-dimensional symmetry. Yet, while *up* is invariant and *down* is invariant, the *up* and *down* are still not unambiguously defined. The left and the right hand are always different (can be differentiated) but a particular hand needs to be associated with upward direction because *up* and *down* as such are not absolute. *Up* and *down* are indeed invariant under mirror symmetry but that by itself does not make *up* absolute to all observers.

Up is not inherently unambiguous. Creation of any motion always results in two objects moving in opposite directions and an agreement needs to be reached which of the two directions is *up* because both the *up* and *down* are always created together. Merely asking for something to go in the *up* direction is not sufficient unless and until the *up* direction is actually defined. Such definition is arbitrary but once the third dimension of space is available, the definition becomes absolute and is, therefore, enforceable. The *left* corkscrew and *right* corkscrew can be unambiguously differentiated from each other but only because these exist in 3D. The left corkscrew can be always identified as the left corkscrew regardless of the viewing angle. By using the *right* corkscrew in the definition, the *up* is now unambiguously defined as the direction in which the right corkscrew is advancing. Further, all observers in the universe are now able to agree on *up* as an unambiguous reference because the direction of motion of a physical body is conserved anywhere in the universe and the *up* reference can be extended to the entire universe. Top and bottom (up and down) of our solar system can be differentiated as you look at the solar system from anywhere around it because the *right* hand specifies *up* under rotation and the right hand fully specifies the angular momentum that is also conserved throughout the universe. A body orbiting another body defines the top of the body it orbits, for the thumb of the right hand defines the top once the fingers are aligned with the orbiting body. Even if two bodies are of the same size and orbit each other, the top is absolutely determined using the right hand. But why stop here? Since the total angular momentum in the universe is zero, there is no such thing as top and bottom for the universe as a whole. The up, down, right, left, forward, or backward are all parameters that arise from motion. While this may seem trivial, consider that the motion must be that of a real thing, for only the direction of motion of a real thing is conserved as it moves through the universe. The direction of motion of a photon, for example, is not conserved, because reflected photons do not leave a "paper trail" that would unambiguously preserve the original direction at the photon's creation.

The law of the conservation of momentum is classified as *vector* law: the speed as well as the direction of a moving object is conserved. However, the vector classification just suddenly appears without any previous explanation regarding the conservation of direction. There is, then, a missing link in the explanation of the law of the conservation of momentum because, while the magnitude of energy is conserved, the direction of motion, which is really a *direction of energy*, is conserved as well. The law of the conservation of direction is presently subsumed because it is not formulated ahead of the law of the conservation of momentum. There will be more to say on the conservation of direction ahead in the chapter on projection.

Two-dimensional symmetry is common among complex organic molecules where mirror, or plane, symmetry yields left-handed and right-handed versions of the same molecule. Left and right-handed versions are atomically identical but differences in spatial geometry give the molecules distinct chemical and physical properties as well. The opposable thumb moves into the third dimension and becomes the key in defining the left and the right hand. Angular momentum also takes the third dimension because angular momentum is perpendicular to the orbiting plane.

● ● ○ ○ ○ ○ ○ **Star Numbers and Operators**

Using two-dimensional symmetry on the atomic level to obtain three-dimensional atomic structures results in one electron having the left-spin state while the second electron will have the reflected or the right-spin state. Helium has twice as many components as hydrogen, with two protons and two electrons. Helium likely has symmetry about a plane and, mapping the spin with left and right states, the second electron in helium can be *included* only if it has opposite spin that results from the reflection about the plane. To show conclusively that helium has symmetry about a plane, the second proton inside the helium core should have its spin reversed, if indeed the spin of both the protons and electrons are reflections about the plane. Plane symmetry works well to grow, by doubling, an entity such as hydrogen into helium. Proteins have unique and intricate structures that exist in 3D. During protein's formation the components fold over in 3D space to complete the structure. Plane symmetry can be applied to advantage to rapidly and without exceptions replicate the most complex 3D structures – where the left-right handedness reversal is the cost of replication.

Pythagorean Tetractys is a symbol consisting of ten dots arranged in a triangular configuration. Tetractys is oftentimes enclosed in a triangle.

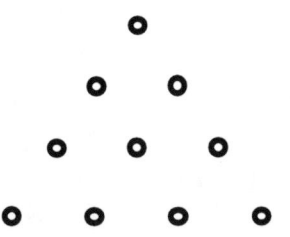

```
         o

       o   o

     o   o   o

   o   o   o   o
```

Pythagorean interpretations of Tetractys are both simple and symbolic. The simple interpretation states that one dot forms a point, two dots form a line, three dots form an area and four dots enclose volume. Tetractys was conceived thousands of years before the high-energy atom-smashing incursion and before the atomic even and odd function classification. There are many structures, both physical and mental, which have preferences for the 0, 1, 2, and 3-dimensional formations. The top dot is about a zero-dimensional entity, which is something that happens to any real object that moves sufficiently far away. The top dot is at times removed or substituted with another symbol.

Quantum Pythagoreans

It is not obvious that nature would favor certain mathematical constructs and one may think nature would not prefer number 1.6 to number 6.1, say. Yet nature must compute if it adheres to or pursues certain purposes. Nature computes if it exists in or creates cyclic behavior. The conservation of momentum calls for two bodies if movement is to arise. Bodies need not be of the same size but there must be exactly two bodies. The integer two, as a number plain and simple, is endowed with a property that makes it very valuable for the purpose of the creation of movement.[9] Similarly, symmetry about an axis is special in that the axis divides the entity having even energy distribution into two equal amounts where symmetries simplify computations. It should then come as no surprise that Pythagoreans thought of maximally symmetrical bodies such as a sphere, cube, tetrahedron, or dodecahedron to be worthy of mathematical and philosophical pursuit.

Twenty-six hundred years ago, Pythagoreans took numerology seriously, but unfortunately the number two did not get good reviews, for number two was also associated with a division of unity. Number two continues to be involved in a division as when the photon's energy is halved to impart energy to two bodies. Yet the two bodies already exist as two bodies and it is the axial symmetry, not the two bodies, which deals with division. The number two, then, is not only numerologically exonerated but also acquires merit for its close association with the creation of movement. Upcoming chapters will show how the unbounded growth of a single body becomes computationally unstable and a split or a spin-off into two bodies then also gives the number two the association with growth. A student could apply Pythagorean knowledge to economics and the creation of wealth. If the student were to major in the military arts he or she could reflect on the symbolic meaning of the edge of a long sword as the axis for division, separation and reduction, as well as a tool for binary (right-wrong, friend-foe) differentiation. A Pythagorean philosopher may contemplate the shape of a triangle while centering a pair of opposites at the tip. A Pythagorean adept with medical

---

[9] Number two is also applied when doubling the object's size, doubling or halving spatial distance, or when doubling frequency. Number two, then, is also synonymous with the octave.

aspirations could use a wand to present an entity with the axis of symmetry such that the entity evenly collects or partitions about the wand.

An additional complication to numerology arises with operations. It may be easy to associate number three with chaos while thinking of three gravitationally interacting bodies of about the same size. Three bodies, however, can also be arranged through a tractable solution with particular values for mass and orbit distances, and then the number three can be associated with a path leading to order.[10]

Tetractys' dot count starts at one, which corresponds to *zero*-dimensional environment. There is, then, an offset by the count of one because a person may think of the start and the finish as two points while another person sees the same situation as one dimension, one path or one link. Because four points define volume, some people associate the number four with material things and visualize it as a cube. Others, and Pythagoreans in particular, visualize materialization as a triangular pyramid, a tetrahedron. The practitioner has no conflict. If the applicable foundation and context are consistent the conclusions are also consistent. Yet, generalized conclusions about numbers require amplification such as the number and the framework of degrees of independence. A numeral, moreover, is the graphical way of describing a number and a numeral may carry additional symbolism as well. Tetractys has ten dots but Tetractys is also a particular member of a triangular sequence of numbers.

Pythagorean numerology is a component of Pythagorean mathematics in much the same way Mother Goose is a component of English literature. Pythagoreans likely used numerology in an introductory fashion when they took simple integers and associated them with larger concepts. Integers are whole numbers that can be even or odd, but numbers also have unique properties that become evident

---

[10] Mythical stories often give gods several roles or powers. Shiva, for example, is the creator as well as the destroyer. Because of Shiva's great musical talent Shiva is the creator of order and also the creator of chaos. Shiva's curriculum vitae of music and dance is most likely related to the family of irrational numbers, which can have harmonizing as well as disturbing qualities. There will be more to say on irrational numbers in the upcoming chapters.

through mathematical operations. If a number divides (parts) a given number without a remainder then the given number is called composite. If no number could divide a given number then the given number was called *incomposite*. While such numbers are today called prime numbers the Pythagoreans thought of these numbers as being incapable of composition. It turns out the incomposite numbers are closely associated with the atomic orbitals the electron cannot occupy. Calling incomposite numbers 'prime' then degrades and possibly even changes the meaning of incomposite numbers because the word 'prime' does not describe certain applications of incomposite numbers. One other collection of operations, operations that transform even numbers into odd numbers and vice versa, remained a curiosity for two and a half millennia before a separate and important branch of mathematics, group theory, arose from it.[11] Other operations are logical as well as numerical. One-to-many relationships and many-to-one relationships, for example, are fundamental constructs in relational databases and in communications, and these are representations of a priority issuing from among many interacting variables.

Division by two is an operation that can be readily applied to all even numbers. If a number can be repeatably divided by two all the way to one then such a number was called evenly even. Today these numbers are called binary numbers. Having 360 degrees in a circle made sense because many numbers can divide into 360 without a fraction left over. The 360 degrees of a circle divide by twelve into 30. 360 days are pretty close to making a full year and the division by twelve is akin to obtaining a month. Having 24 hours in a day also means we calculate earth's rotation as 15 degrees per hour or one degree every four minutes.

---

[11] It is said Confucius was fond of I Ching, a divinatory custom that deals with the change of Yin into Yang and vice versa. Yin is represented as two bars (even) while Yang is a single bar (odd). I Ching also pays attention to changes from subordinated to dominating position (below and above) and vice versa. Confucius lived about the same time as Pythagoras, 550 BC. Buddha is a contemporary of Pythagoras and Confucius as well. Maya first appear around 300 BC.

● ● ○ ○ ○ ○ ○   Star Numbers and Operators

Geometry and Arithmetic

The angle of 120 degrees is constructible exactly with only a compass. Any angle can be readily halved using a compass, and exact angles of 180, 90, 45 and 30 degrees are easy to obtain as well. Having a circle consisting of 100 degrees, for example, arithmetic division by three, six, eight, or twelve would not yield an integer. Pythagoreans called a number that can be divided by many other numbers without a remainder the abundant number – and fifteen different numbers can divide into 360. However, selecting a number with many divisors to represent degrees in a circle does not mean that arithmetic and geometric divisions are equal. Number 364 is divisible by 52 as well as by 7, but a circle cannot be divided into seven equal parts using a compass and straightedge. There exists a compromise between arithmetic and the geometry of the circle. Arithmetic allows us to pick any number for the quantity of degrees in a circle and it then makes sense to pick a number that has an abundant number of divisors. Geometry, on the other hand, lives with the π that strictly ties the circle to the transcendental number that is π. The transcendental number, just as the irrational number, has an infinite mantissa. As the transcendental number is being formed, ever-smaller components are added toward completing the number – but additions must accommodate an infinite number of components. Digits of the mantissa express the transcendental number but there is no ending or final digit. Further, any transcendental or irrational number that is divided by any real number retains its infinite mantissa, and so every possible degree in a circle is not a whole or a real number. Not all is lost, however, because geometry allows the division of a circle into particular and *exact* number of segments.

The enigmatic result is that arithmetic cannot divide a circle into real and exact segments but geometry can. The most appropriate application is with the atomic orbital. An atomic electron is also a wave that acquires stability when it closes upon itself as it wraps around the atomic core in the exact multiple of its wavelength. Arithmetically, the division of a circle into seven segments can be carried out to a large number of decimal places and this computing procedure would be similar if another number such as five were used in the division of a circle. Yet, five wavelengths will fit the circle exactly but seven wavelengths never

Quantum Pythagoreans

will. Geometrically it is possible to divide one irrational number by another irrational number and obtain a real number – but the computer cannot hold the irrational number in its entirety. One fifth of a circle going over the periphery will meet exactly one fifth of a circle going in a straight distance of a cord. But the same property does not hold for the number seven. The number three geometrically divides a circle of 360 degrees exactly into three angles of 120 degrees each, but the number three cannot divide the remaining 120-degree angle exactly. Because the atom's existence is sustained through computing, there will never be an atomic orbital that consists of seven or nine wavelengths. There will be more to say on the geometric creation of irrational numbers in the framework of the workings of the Great Pyramid.

Compass and straightedge are representative of tools that can solve quadratic equations. It is likely that on the macro scale of the universe the quadratic methods are the generally available methods, particularly since the real parameters of large planetary masses cannot readily transform. Quadratic equations have two solutions. On the analytical level, both solutions can be exchanged – and one solution used in place of the other – by using symmetry about a point (rotation) *or* using symmetry about a line (reflection). The exchange of solutions from quadratic equations is then commutative, which means that the order in which exchanges are applied does not upset the before-and-after validity of solutions. This property can be interpreted as follows: "Methods can be selected out of the hat and either of these methods – rotational or translational – will yield correct solutions. God likes methods that can be picked out of the hat to show that the methods were arrived at without prejudice." Rotational and translational processes are facilitated with compass and straightedge. Solutions of cubic (third) and quartic (fourth) order equations map into tetrahedron and a cube, respectively, but only a subset of these solutions has commutative property when the methods are exchanged under symmetry. Other Platonic solids are also placeholders for other solutions, which are then moved about in these highly symmetrical settings. Mapping and exchange of methods is also fundamental in the mathematical group theory, which, among other things, deals with invariance and integrity during transformation.

Quadratic equations have two solutions and in case such as $x^2 = 25$, solutions are +5 and -5. It is tempting to speculate that

43

Pythagoreans were postulating the invisible "counter-earth" on the opposite side of the sun when they were working quadratic relationships and realized there exists more than one solution. Aristotle pokes fun at the Pythagorean notion of a counter-earth, contributing it to Pythagorean desire to find an extra planet for a total of ten planets, in agreement with Tetractys.

Pythagorean mathematics deals with the understanding and application of mathematical operations as analogous to the operations of nature. While addition appears trivial, its counterpart in physics is superposition, which finds applications with light as well as gravitation. Multiplication may be thought of as simple leveraging, yet the most unique aspect of multiplication and squaring is its tie-in to the physical aspect of force and to the creation of force. Force is the most difficult, if not impossible, parameter to define directly and is usually defined by its consequences such as the ability to stretch a spring, speed up an object, or smash things up. At times force is combined with the parameters of mass and distance to get energy and make more sense out of it.

Tractability Revisited

When three or more bodies interact gravitationally the mathematics seems straightforward. There are no inherent barriers that would hide the knowledge of such variables as distance, mass size and velocity. Knowing these variables, gravitational forces between any two of the three bodies can be readily computed and the net force on any one body can then be computed by addition just as readily. However, when all equations are written down and a solution sought for the force a particular body is subject to, the one unexpected result is that there exists no mathematical solution. Three-body configuration has no *general* mathematical solution and the absence of a solution offers a good interim definition of chaos. A particular body is subject to infinitely varying changes in force magnitude and direction, and the absence of a mathematical equation does not allow force to be, by definition, deterministic. If there were an equation describing the force, the path a body takes would then be tractable as well even if it were a shape of, say, a corkscrew that gets tighter or looser or stays about the same.

Intuitively, if a body's path repeats then there should be a solution and indeed chaos does not have a repeating period as one of its characteristics.

In the absence of a formula the path of three bodies can still be simulated on a computer but the simulation does not hold true for long. Starting with the exact knowledge of all bodies' position, mass, velocity and force, the computer uses a time increment to calculate the next body position, for taking a time step is the only way of simulating the system without the equation. Using time increments introduces a computational error into the result that can only get worse as the simulation proceeds. The time increment can be shortened to obtain better accuracy but this also results in increased computing time. The limitation and trade-offs of three-body computer simulations surface when the perfect three-body simulation is reached at the expense of infinite computing time. The simulation of three and more bodies is, in general, intractable.

In the search of a tractable solution to multi-body problems the issues are becoming clearer, inasmuch the operation of a reduction is to work on all possible paths with the result of overlaying all possible paths to some path – be it linear, curvilinear, or oscillating – that would then repeat. Because the one-body system is also tractable, the process of path reduction deals with changing multi-body chaotic systems or potentially chaotic systems into a compound two-body or one-body system. One of the one-body solutions is the accumulation of bodies into one larger body. Gravitational force between bodies attracts bodies together where the resulting accumulation is in accord with a one-body solution. Gravitational linear attraction is – and can be visualized as – a tractable one-body outcome from among many bodies. Yet, the linear attraction is not the only way of obtaining a tractable solution. Angular momentum keeps bodies in orbit and deals with two-body systems. A two-body periodic solution requires an orbit where two bodies do not collide, and the underlying ways-and-means then call for the mechanism that creates gravitational spin. Theoretically, the accumulation of matter into a single body would or could create a single and a straightforward solution. There are, however, limits on the size of bodies because atomic stability begins to weaken as mass increases. The limits will be described in chapters on organization. A solid one-body solution is not a universal solution and neither is a gaseous one-body solution. Geometries are an integral part of the solution and *orbits* are necessary to further increase and grow the universe.

Before the concept of chaos reduction becomes overly mysterious and require mystics of mystical proportions, let's assert that the most practical way of putting Oss together is to start with one body and then spin off additional bodies as necessary and in context where chaos does not develop in the first place. Size and distance of each new body will require much computation that results in the arrangement minimizing the propensity for three-body chaotic interactions. Reduction in chaos, then, is about creating body sizes and distances where mathematical solutions can be had.

Ratios Play Musical Orbits

Three or more gravitationally interacting bodies are, in general, chaotic. Our solar system, or Oss, has more than two bodies but Oss is not chaotic because the massive central sun is capable of forming a two-body system with each planet. Functionally, Oss is a multiple or a compound two-body system where each two-body system is computable on its own, and there is some aspect of independence inherent in this configuration because planets have "their own" orbit. Johannes Kepler was able to fit a mathematical equation such that a particular planetary orbit period depends on the parameters of the sun and the planet – while the parameters of other planets are not needed. There are planet-to-planet gravitational influences but these appear small enough not to disturb planets much in their independently computable two-body orbits. However, even small influences can add up to significant influences if the interaction is periodic because in the periodic environment the influences become additive. Planets are, by definition, periodic. All planets, then, should be found in a state where their mutual influences even out over time. A rational relationship among orbit periods will accomplish that. Because rational fractions always repeat, then a finite, fixed, and predictable amount of overtones continue to repeat. If the Earth's orbit were found to be increasing by few seconds each year, a rational relationship will in time guarantee a repetition and Earth's orbit will start to decrease – and then increase once again because the fractions

Quantum Pythagoreans

of rational periods are repetitive. Further yet, some rational orbit fractions may be better than others.

Orbits of Venus and Earth are interlocked in a ratio. For every five Earth orbits Venus completes eight orbits. Orbits of these two planets are not independent of each other but this is not a source of orbital problems because the orbital interlock of Venus and Earth allows both planets to become effectively a single planet in one orbit. The resulting planet is moving close to the midpoint between Venus and Earth because both planets are of about the same size. From the Sun's point of view the positions of both Venus and Earth periodically occupy the same two spots on the solar plane and the merged planet is then also in a spot it occupied before. For this merged planet the aphelion period is the time for Venus to catch up with Earth again, which takes 1.6 Earth years. Every time Venus catches up with Earth the merged planet reaches its aphelion, the farthest distance from the center of the Sun. Orbital interlock merges both planets yielding one highly elliptical orbit that rotates and repeats every eight years – that is, the merged planet passes the same starting point every eight years. During the eight years the merged planet reaches aphelion five times, tracing a five-pointed star with 1.6 years between points.

Mayan astronomers first discovered the Venus-Earth periodic relationship, possibly more than a thousand years ago.[12] Planetary interlock similar to Venus and Earth also exists for the orbits of Neptune and Pluto, and in both cases the orbital interlock merges two slightly elliptical orbits into a single orbit. The 8/5 orbital ratio in the Venus-Earth case results in the five-pointed star shape orbit of the merged Venus-Earth planet. The 3/2 orbital ratio in the Neptune-Pluto case results in the two-pointed star or lentil shape orbit of the merged

---

[12] Five Venus-Earth aphelions bring Venus and Earth to the same points on the solar plane and the number five is prominent in the Mayan numerical system. Mayans use symbols for zero, one, and five in a summing sequence that becomes a compound vertical numeral from zero to nineteen. Positional notation with base 20, also called vigesimal notation, is then applied to express any count. Mayans did not use fractions or the vigesimal point. There is an exception to Mayan count that has a counterpart in our system. It is analogous to our counting seconds up to 59 rather than 99.

● ● ○ ○ ○ ○ ○   **Star Numbers and Operators**

Neptune-Pluto planet.[13] The five pointed Venus-Earth combined orbit makes a tight curlicue around the sun while the two pointed combined Neptune-Pluto orbit makes a large curlicue around the sun before making the second point of the two pointed star. After two Pluto orbits the merged planet traces its own complete orbit resembling a catseye.

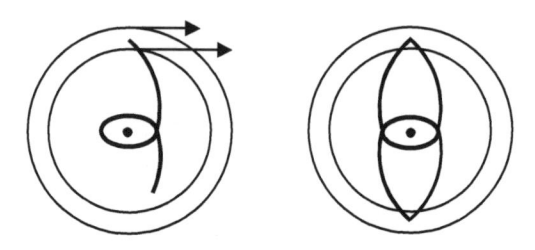

Single orbit from Neptune-Pluto merged planet traces Catseye orbit

Venus-Earth orbital interlock allows merging of these two planets into one where the resulting body is computable with respect to the Sun – for any two bodies are always computable. A potential Sun-Venus-Earth chaotic trio becomes a computable duo. The Venus-Earth merged planet moves around the Sun and transcribes its aphelion points in an interesting sequence. Every two points are skipped as the merged planet's orbit moves from the first point to fourth to second to fifth to third and back to the first point. The aphelion points are not made sequentially along the orbital circumference and the trace of the merged planet is the pentagram. Neither the pentagram nor the catseye orbits can be observed but they can be visualized or computed. Merged planets do exist because the Sun actually gravitationally feels the merged planet in the pentagram orbit and another merged planet in the catseye orbit. Since strict visual proof of existence is not available, interest in the pentagram shifted to sciences that welcome a more intuitive or computational

---

[13] The closest symbol to the two pointed lentil shape orbit is *Yoni* or *Vesica Piscis*, which results from the intersection of two circles. Sacred geometry uses a straightedge and a compass to create shapes having certain proportions, shapes and symmetries. Merging of two planets having two interlocked orbits results in an orbit that is always continuous – that is, the points associated with merged orbits may be tight and sharp but they are always rounded.

approach – the idea being that nature as well as a person's mind must and do compute.

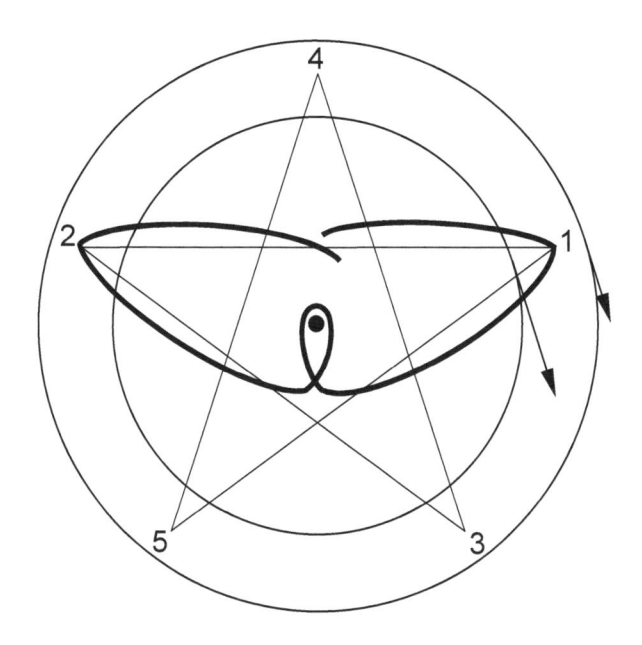

Pentagram orbit of merged Venus-Earth planet
Orbits to scale

What distinguishes the pentagram from a generic five-pointed star is the order of drawing the points. Pythagorean tradition calls the pentagram the pentalpha because five A's also make this symbol unique. Lower case alpha contains a curlicue as well. Many different associations were made with the pentagram, and the computing framework sees the pentagram as a stabilizing orbit that happens when two orbits interlock in the ratio of 8/5. Being in the solar plane the pentagram's orientation in and of itself does not have an appreciable social association – good or evil – and in fact the pentagram slowly rotates in the solar plane because the 1.6 years between actual aphelions is very close to 1.6 but is not exactly 1.6. The pentagram's rotation does not change the pentagram's associations and, therefore, the pentagram's rotation cannot corrupt its symbolism. Two orbits are necessary for orbital pentagram creation and one may muse over the custom of having two rings as the symbol of interlocking or bonding a couple in a marriage. In the Pythagorean

● ● ○ ○ ○ ○ ○   Star Numbers and Operators

tradition the pentagram is enclosed in one ring, which is then made into a piece of jewelry – a ring. From Kepler's orbit laws the ratio of two radii of pentagram-producing rings can be computed. The inner ring is obtained by multiplying the outer ring by 0.731.[14]

Earth, moreover, has an orbital interlock with Mars in a ratio of 15/8. For every 15 full Earth orbits Mars makes full 8 orbits. Earth and Mars then also form a merged planet that makes an eight-pointed star. Overall, then, Earth forms one interlock with Venus and another interlock with Mars and each of these duos make one computable orbit from the sun's perspective.

How many real orbits Oss could have is a complex issue. Interlocks continue to reduce the orbit count through computational means and in the end there is but one merged planet with one merged orbit. This method works from the solar perspective and holds for any solar system. Observations of some other solar systems indeed result in the detection of a net wobble on the central sun. Presently, however, national agencies attribute such wobble to one planet in one orbit but it is just a net merged planet that is being inferred. Announcements of a planet discovery at another solar system unfortunately refer only to the net effect of all planets in a particular solar system. A solar system could be well balanced with many planets and then the overall wobble may be so small it is not detected with present day instruments.

For the Earth-centered observer the orbits of planets do not merge because the context of the observer is different. From the Sun's perspective the moon is so small it merges with Earth – but the moon exerts the greatest influence upon the Earth-centered observer. Kepler's astrological notes indicate that some star configurations were 'more harmonious than others,' but he did not explain why. Kepler used the word 'star' in a context where the observer and two of a planet's positions, (our moon is included among planetary bodies), made evenly

---

[14] Applying Kepler's equation relating the orbit time to the radius: $T^2 = k \cdot R^3$, the ratio of radii are computed by substituting the orbit time T for both planets. For inner to outer radius ratio, $5^2/8^2 = (R_i/R_o)^3 = 0.731$. The arithmetic easily dispenses with the cube root, which is a difficult problem for geometry. Taking a cube root is inherent in the 'doubling of the cube' problem of ancient Greeks.

spaced 'points' of regular stars. If the angle between a planet, earth, and moon made a 90-degree angle then Kepler was dealing with a four pointed 'square star' because 90 degrees were folded into a square without a remainder. When applying the geometric means of the compass and straightedge, the angular position between two planets and the observer on earth is not always possible to construct exactly by geometric means, for no orbit (a whole circle) is divisible by seven, for example. There is, then, some rationale for seeing the constructible stars as being 'more harmonious.' While it may appear that the phrase *geometric means* is in some ways arbitrary and limiting, it may be helpful to equate this phrase with the ability to construct waves. Applying geometric means, then, is asking the person not only to use a compass and a straightedge to divide a circle, but also to construct the actual distances in multiples that benefit and cater to the properties of waves. The waves need exact equal distances in integer multiples in order to form a *closure* in a geometry of a circle – a geometry that is symmetrical about a point. A compass is exactly the instrument for the creation of symmetry about a point. Later on Carl Gauss discovered the theorem that describes which polygons could and could not be constructed with compass and straightedge.[15] A case can be advanced that the constructible polygons are, on account of their exactness, more harmonious. In Kepler's case, the angle that equals one seventh of a circle could be stressful for lack of the actual standing wave – or for the presence of disharmonious waves. All said, the rational relationship among planetary orbits could add additional explanatory dimension to planetary harmony.

Should two planets interlock in the orbital ratio of 5/3, the resulting merged planet would orbit in a three-pointed star topology. Although such interlocking arrangement does not exist here at Oss, it would make for yet another stabilizing planetary arrangement. Kepler's

---

[15] The regular polygon constructability rule is easy: A circle can be divided exactly using a compass and a straightedge into 2, 3, 5, 15, and 17-sided polygons. Since any angle can be halved exactly, the 2-sided polygon will spawn a family of 4, 8, 16, ... sided regular polygons. Similarly, 3, 5, 15, and 17-sided polygons will *each* spawn a family of their own by doubling.

equation could also be used to calculate individual planetary orbits that would make the orbital ratio 5/3.

Pythagoras is credited with the invention of notes that step through the musical octave. A new octave happens every time a musical string's length is halved and in consequence the string's vibrating frequency doubles. The individual notes within the octave were chosen, according to lore, for their harmonious and illness-curing effects.[16] During the Renaissance the planets were placed on "musical spheres." The orbit period of a planet keeps on repeating and that is also associated with a particular frequency a planet has. Alchemists of the Renaissance spoke of the music of the heavenly spheres and they gave every planet a note, but they did not go as far as to apply the interlocked planetary orbits to a particular *dual* note. The following table shows the notes of the octave and their vibrating frequency *ratio*. The lowest C note is normalized at one, and each note then has a vibrating ratio referenced to the lowest C note. Two octaves are shown.

C	D	E	F	G	A	B	C'	D'	E'	F'	G'	A'	B'	C''
1/1	9/8	5/4	4/3	3/2	5/3	15/8	2/1	9/4	5/2	8/3	3/1	10/3	15/4	4/1

The ratio 3/2 of the Neptune-Pluto orbits results in the catseye orbit and corresponds to G *and* C notes because both notes are needed to produce 3/2 ratio. E' and A notes also sound with the ratio of 3/2 where E' is the E note of the next higher octave. Take the higher frequency and divide it by the lower frequency to obtain ratios. The orbits ratio of 4/3 results in a three-pointed star orbit (if it were to exist in Oss) while corresponding to F and C notes. The ratio 8/5 of the Venus-Earth orbits results in the pentagram orbit and notes C' and E provide the ratio of 8/5. An octave can start at any note and starts at C by convention. Starting at

---

[16] Individual notes in the octave are ratios of whole numbers that are rational numbers, and such ratios have the property of a repeating fraction, be it decimal or any other base. Any number or a sequence of numbers begins to repeat sooner or later. 1.6̲00000… 1.6̲66666… and 1.4̲81481481… are all rational numbers that in these cases are results of ratios from the Pythagorean, or Western, octave.

Quantum Pythagoreans

any note and ending at the same note of the next higher octave always sounds with the vibration ratio of 2/1. To emulate the interlocked Oss orbits musically the proper ratio can be found within any octave starting at any note. Playing G and C notes within any octave harmonize with the Neptune-Pluto orbit. Harmonizing matches the instrument's vibrating frequency ratio with the planets' orbiting frequency ratio. Playing C' and E notes within any octave (E is lower and C' is higher) harmonizes with the Venus-Earth orbit. Tones produced with C and B notes have ratio of 15/8 and harmonize with the eight-pointed star, which is the Earth-Mars orbit ratio. The following table shows the ratio between two notes when sliding across an additional octave. Notes in the left column are the lower of the two.

	C	D	E	F	G	A	B	C'	D'	E'	F'	G'	A'	B'
C	1/1	9/8	5/4	4/3	**3/2**	5/3	**15/8**	2/1						
D		1/1	10/9	32/27	4/3	40/27	15/9	16/9	2/1					
E			1/1	16/15	6/5	4/3	**3/2**	**8/5**	9/5	2/1				
F				1/1	9/8	5/4	45/32	**3/2**	27/16	**15/8**	2/1			
G					1/1	10/9	5/4	4/3	**3/2**	5/3	16/9	2/1		
A						1/1	9/8	6/5	27/20	**3/2**	**8/5**	9/5	2/1	
B							1/1	16/15	6/5	4/3	64/45	8/5	16/9	2/1

Starting with the higher note, Venus-Earth orbit ratio of **8/5** can be expressed as dual notes C'-E, F'-A, or G'-B. Mars-Earth orbit ratio **15/8** tones can be expressed as dual notes B-C or E'-F. Neptune-Pluto orbit ratio **3/2** tones can be expressed as dual notes G-C, B-E, C'-F, D'-G, or E'-A. All ratios for a particular planetary pair are to be found on a particular diagonal.

Several other interlocked orbits can be produced with the notes of the octave. In addition to the ratio of 5/3 that would make a three pointed star and correspond to A and C notes, the ratio of 4/3 would also

53

● ● ○ ○ ○ ○ ○   **Star Numbers and Operators**

produce a three pointed star in tune with C' and G notes. The ratio of 5/4 belongs to a four-pointed star harmonizing with E and C notes. G and E notes have the ratio of 6/5 that would make a five-pointed star but it would not be the pentagram as each point of this five-pointed star is traced in incrementing sequence. C' and D notes with the ratio of 16/9 corresponds to a nine pointed star where all odd points are traced first followed by all even points. Notes D' and A result in the ratio of 27/20 that produces a twenty pointed star by making every seventh point until all points are made. This star completes in a sequence: 1, 8, 15, 2, 9, 16, 3, 10, 17, 4, 11, 18, 5, 12, 19, 6, 13, 20, 7, 14, and back to 1.

Many ratios in the two-octave progression match terms of the series **(n+1)/n**: 2/1, 3/2, 4/3, 5/4, 6/5, 9/8, 10/9, and 16/15 – which are the ratios of Pythagorean triangular numbers. There is one ratio that is not a ratio from the musical octaves but it is used by two of Jupiter's moons: 7/3. Note that this ratio spans more than one octave. Within one octave, this ratio would be 7/6, and in agreement with the triangular series **(n+1)/n.** Ratio 7/6 would trace a six-pointed star orbit, but this ratio does not appear to exist among planets or moons of Oss. Numbers 7, 9 and 11 are the only numbers between 6 and 12 that would allow a construction of the six-pointed star within a single octave. Numbers 7, 9 and 11 cannot divide the orbit period (circle) into equal segments through geometric means.

Any two orbits can be constructed into a multi-pointed star. Initially, it is not necessary to be concerned with practical limitations and stars can be constructed with the following simple formula: Make any orbit ratio in the format **(x+y)/x**. The integer **x** is the number of points the star will have. The integer **y** is the number of the point that is the next point traced. When **y** is **1**, points are traced in clockwise order and the star becomes the regular polygon. In the case of the 6/5 ratio, for example, a pentagon is created. In cases of 8/5 or 7/5, a pentagram is created, for tracing every third (y = 3) or every other (y = 2) point of the five-pointed star results in the pentagram. A similar result is for ratios 5/3 and 4/3, which create three-pointed stars – one in the counterclockwise order and the other in the clockwise order. An even-pointed star cannot happen by tracing every other point (y = 2), because only half of the points would be traced. Therefore, the 10/8 ratio does not result in the eight-pointed star but results in a four-pointed star that is in accord with

Quantum Pythagoreans

the 5/4 ratio. In the case of the 13/8 ratio or (8+5)/8, the eight pointed star is created by making every fifth point, as shown below on right.

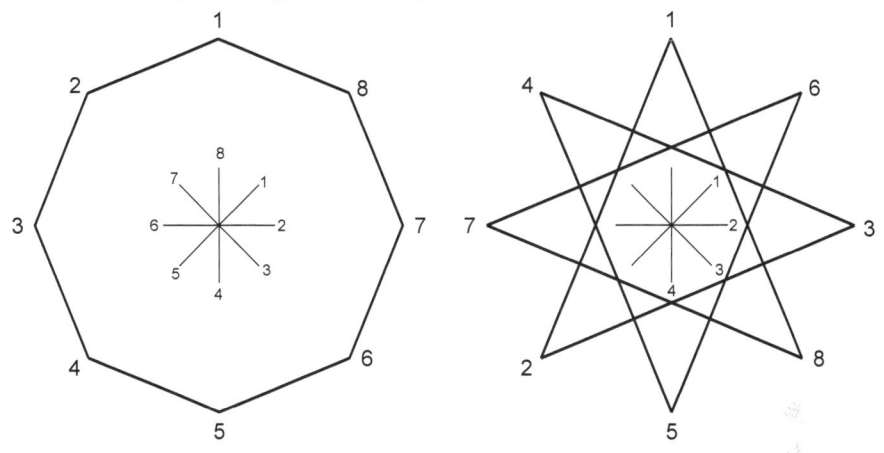

Octagon Star from 15/8 Orbit Ratio

Octagram Star from 13/8 Orbit Ratio

    The star identical to the 13/8 orbit ratio star can be made with the 11/8 ratio, except that the order of tracing the points would be different and drawn on paper in order 8 (top), 3, 6, 1, 4, 7, 2, 5, and back to 8. Reviewing the Earth-Mars orbits in the 15/8 ratio, the resulting eight-pointed star is the regular polygon of the octagon, as every 7th point is traced – shown above on left. As a final check, the 12/8 orbits ratio will trace a two-pointed star in the general north-south direction. 12/8 ratio can be reduced to 6/4 and also to 3/2 ratio, all of which produce the two-pointed star of the combined Neptune-Pluto orbit.

    Mathematicians might recognize the formula **(x+y)/x** as a form of modulo arithmetic. This is indeed the case but blindly following modulo rules will not recognize the 10/8 ratio in the appropriate 5/4 format. You may and should adjust the math rules, but also consider that **x** or **x+y** cannot be equal to 7, for example, because the additional rule needs to include the constructability of Gauss' regular polygons. Both the numerator and the denominator must be constructible through geometric means. In the applications, the construction of the octagram in ratios 13/8 or 11/8 will not be harmonious. Finally, the appearance of number 9 in the ratios of musical notes may seem an enigma if we presume that all ratios of all notes comprising our octave "must be" harmonious. The ratio of notes having 9 in the numerator *or* in the denominator will likely not sound harmonious. Indeed, playing D-C or

●●○○○○○ **Star Numbers and Operators**

E-D together as dual tones do not sound harmonious. Why Pythagoras included notes in the octave forming a disharmonious 9/8 ratio from D-C but did not include notes forming another disharmonious ratio of 7/6? It is because notes relaying on thirds sound harmonious with all other notes except those notes that also have a three in their ratio. For example, playing A with all notes *except* G will be harmonious. Both A and G have the number three in their fraction that will result in nine when notes A and G are played together – and so playing A and G together is disharmonious. A and G each bring thirds that individually divide circle exactly and so are harmonious with other notes. Harmonious ratios will happen only if the points of the stars in the numerator and the denominator are *both* constructible exactly through geometric means. These numbers may be called circumpositional, for they fit exactly around the circle.

The following circumpositional numbers from 2 to 20 divide the circle exactly and using any of these numbers in the numerator and denominator will sound harmonious together: 2, 3, 4, 5, 6, 8, 10, 12, 15, 16, 17, and 20. Circumpositional numbers may find applications in atomic construction.

The following table compares the orbit ratios of some of Oss planets and moons – both as expected from the ratios of the musical octave and as measured. One to two-octave reach operates, or "plays," between Earth and the rest of the terrestrial planets.

Quantum Pythagoreans

	Musical Ratio	Actual Ratio as measured
Earth and		
Mercury	1/4 = 0.250	0.241
Venus	5/8 = 0.625	0.615
Mars	15/8 = 1.875	1.881
Mars' **Phobos** and		
Deimos	1/4 = 0.250	0.253
Jupiter's **Io** and		
Ganymede	1/4 = 0.250	0.247
Europa	1/2 = 0.500	0.498
Pluto and		
Neptune	2/3 = 0.666	0.665

It is easy to guess that the relative size of mass and the sun-planet separating distances have something to do with avoiding three-body chaotic configurations. Orbital interlock helps as well because two planetary bodies in effect merge into one body and the body count decreases. In addition, the angular momentum of Oss is close to maximum because all planets are in one plane and orbit in the same direction. Being in one plane and orbiting in the same direction allows each planet's angular momentum to add consistently in magnitude and in direction thus forming the Oss primary angular momentum. It is also not a leap of faith to say that the Oss configuration is a particular solution of a multi-body problem where chaos and collisions are not desired.

Extra Moon

Earth has a pretty moon some 240 thousand miles away. This moon orbits around the Earth every 27⅓ days, the proper number to use for the orbit time period. After this orbit time the moon regains the same Earth-moon position but the moon does not look the same from Earth because both the Earth and the moon are now about one month forward in their orbit around the sun. For the moon to again have the new crescent shape takes just a bit longer: 29 days, 12 hours, 44 minutes, and 2.8 seconds, a period closely watched and made more accurate since antiquity.

Technically, it is now easy to create a new, second moon and place it in the orbit around Earth using mathematical tools in place. All other constraints are qualitative in nature but the most important one is the long-term stability. For stability reasons the orbit ratios of the two moons will be in the ratios of the musical octave. Multiple moons also orbit Mars and Jupiter. These moons have orbital ratios equivalent to two octaves – that is 1/4 or 4/1, depending on which orbit period is placed in the numerator. By imitation, our new extra moon would have close to a one-week orbit period if placed in the inner orbit and closer to the Earth. Using two-octave separation gives the two moons a rational number to share, and but finite harmonics to deal with. If orbit periods were made with an irrational number, it would mean that the disturbances would keep adding up with each new orbit.

Having settled on the ratio of the two periods, the ratio of the two moons radii can be calculated as well, and with it the actual moon orbit insertion. Kepler's orbit equation relates the square of the orbit time period to the cube of the orbit radii. The two-octave orbit period separation results in a new orbit distance from Earth that is about 40% of the present moon's orbit. Now we can assign mass to the new moon such that the new moon's gravitational force does not upset the existing tides much. One twentieth of the mass of the present moon will give the new moon pull of about 32% of the existing moon. Tides due to the inclusion of the extra moon would become quite complex but the new moon's orbit speed is so rapid that the tides' peaks would not be in excess of existing peaks. Overall, tide excursions would decrease because most of the times the two moons' gravitational pull would be in oppositions and tides peaks would on average decrease. The idea of picking the inner orbit for the second Moon is to make the new moon easier to access and utilize it as a more convenient staging base. A 7-day spin on the new moon will make it stationary with respect to Earth, too. The second Moon could then become the ideal Earth's observatory. The only unknown is the effect the second moon would have on the Earth's atmosphere. A single moon may keep the currents churning but two moons may localize the weather considerably.

Irrational and Divine

Rationing is about partitioning of whole numbers that is related to the ability of all real things such as planets to form interlocking orbits. Having an integer in the numerator as well as in the denominator results in a rational number. By dividing the numerator by the denominator, a rational number becomes a real number. All integers and rational numbers are *real* numbers and real numbers play a role in forming electrons' orbitals in atoms as well. All rational numbers such as 8/7 or 12/101 can be reduced into a real number and transformed back into a rational number. Any rational number always fits exactly into a finite mantissa because some part of their decimal fraction sooner or later begins to repeat, and a rational number can always be stored and retrieved from a computer without loss in fidelity. Real numbers can then also be called the finite or exact numbers.

It is said Pythagoras discovered *irrational* numbers. Pythagoreans of the founding school held irrational numbers to a mystical level because irrational numbers were difficult to express verbally. Irrational numbers are created with integers but not through their mutual division. An intuitive approach is called for when, for example, the irrational number such as the square root of two is applied to double the area of any square exactly, but *only* if the square root of two is not reduced into a real number.[17]

---

[17] It is easy to double the area of a square – and do it exactly through geometric means. Doubling of a volume of a cube, however, becomes a difficult (some say impossible) task for geometry's straight edge and a compass. There is indeed a change in computational modality when moving from two to three dimensions.

● ● ○ ○ ○ ○ ○   **Star Numbers and Operators**

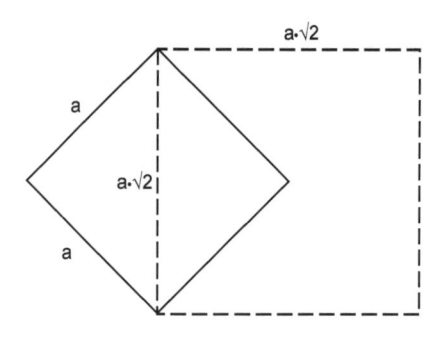

Area of dashed square is exactly twice of solid square

Presently it is easy to explain the difference between exact and precise. If you are using a computer and obtain the value of the square root of two, you will end up with a truncated number because the storage of each and every number has finite length. Subsequently squaring such a real number will not result in quantity two but in a number slightly less than two. The finite length of a number in a computer is called mantissa, which expresses the *precision* of a number. In a calculator, the precision is usually the same as the number of digits on display. Most, if not all, mathematicians do not reduce irrational numbers and perform computations while keeping irrational numbers as a particular and separate class of numbers. It would also be appropriate to classify irrational numbers as irreducible because the act of reduction forever blocks the reconstruction of the number back to its full original value – that is, the transformation of an irrational number into a real number is not reversible. Computers in use today, however, reduce all numbers into real numbers. In any case, it is nice to know that in nature some relationships have a mechanism that is exact and, for example, since the object's moving energy is proportional to the square of its velocity, the object's moving energy will be conserved exactly even if the velocity results from a solution to a quadratic equation.

Irrational numbers have the faculty to carry within themselves the infinity of values that the real number could accommodate only by having an infinite number of decimal places – a practical as well as a real impossibility. Here then is a number that the real number could only approximate.[18] Having an infinitely large mantissa in a number is not possible and now it becomes apparent that the creation of irrational numbers can happen only through geometries. Here now is a class of numbers that results from everyday real numbers but through the operation *other than* the usual addition, subtraction, multiplication, or division. Irrational numbers arise when spatial direction is their construct and the Pythagorean theorem their relational operator. At the heart of the Pythagorean theorem is the square root, and *taking root* is the operation that is unique to geometry. Square root may seem a simple reversal of multiplying by the same number but taking root is not something that is necessary when working with numbers that represent real things. Indeed, having five rows of apples in five columns uses squaring but taking a square root of 25 apples does not have much utility even if the cook wants to make five apple pies. Greek-speaking scholars pursued geometry with passion and geometry did not relinquish its position as queen of mathematics for 23 hundred years and not until Kepler's mathematics became capable of actually predicting a planet's position along its planetary orbit.

A proposition can be made that Kepler's numerical methods for the first time exposed a need for a calculating machine that would be able to predict planetary positions. Yet, when it comes to irrational numbers the present day computer always creates irrational numbers through an

---

[18] Some mathematicians classify irrational numbers as real numbers but that certainly is not the case because it treats the infinite mantissa of an irrational number as a real number. Infinite mantissa, however, cannot be stored in a computer and consequently must be truncated. Truncation stops the irrational number's creation process. Making irrational numbers the same as real numbers seems to satisfy a need for simplifying reduction, but classifying irrational numbers as real numbers is just an assertion that lacks supporting justification. Real numbers represent real things and real things are always finite. Some mathematicians claim the irrational number is equal to the real number because the irrational number can be expressed to as many decimal places as we want. Not so. Every computer will always run out of memory – just as any mathematician will always run out of paper – when expressing an irrational number. By truncation, the irrational number transforms from irrational number into a real number and such transformation is not reversible because the irrational number cannot be restored once it is truncated. Irrational numbers are not real numbers and are in a class of its own.

infinite process that sooner or later must be halted. Geometrically, however, irrational numbers are readily constructed using real numbers and in *finite* time. The geometric construction of irrational numbers is then tractable. This is not just a case of semantics. Instead, it points to a fundamental shortcoming of the present day computer that is limited in its use to but multiply, divide, add, and subtract to accomplish numerical processing. The present day computer has a square root as one of its functions but it is only an approximation of a geometrical square root. The computer's square root substitutes arithmetic operations in the approximation of a square root, and that is the reason the computer's square root is an infinitely long process. Geometrically, however, the right angle construction that yields two points on the diagonal also results in the implementation of the square root because the distance between the two points in space *is* the square root. The finite time creation of an irrational number as a distance between two points is the naturally finite outcome from the geometry of angles, lengths, areas, and spatial separation distances. A case can be made that spatial distances and spatial relationships allow a new group of numbers to arise. The driving parameter is tractability, for tractable computing allows numbers to manifest themselves in nature.

There is also a group of irrational numbers called transcendental numbers. In addition to the ancient duo of the straightedge and compass, transcendental numbers require specialized geometry for their construction – the pyramid. The most famous of the transcendental numbers is π. Transcendental numbers are in the irrational family of numbers, for transcendental numbers cannot be made through rationing. Yet the rationing property is but one property that categorizes numbers. Both the transcendental and irrational numbers share the property that their mantissa never repeats. The property of tractability, moreover, has priority over rationing because tractability allows numbers *to become*. Transcendental numbers do have the property of not being created by rationing, but the infinite addition mechanism that is necessary for their creation puts transcendental numbers in their own separate category. The addition of an infinite number of terms also puts high demands on tractability, for, in our real world, the addition of an infinite number of terms is not tractable. There is one consideration, however, that makes all the difference: The addition mechanism (superposition) of the quantum mechanical wavefunction happens without delay. Mathematically, Gottfried Leibniz identified transcendental numbers through their infinite addition mechanism and even came up with a particular infinite series that converges toward π.

Pythagoreans also called irrational numbers the unspeakables, for it is not possible to describe such numbers by words only, or, if you prefer, you would need an infinite number of words to describe them. Today, a Pythagorean could well call irrational numbers the unwriteables, because no device can print out an endless number. If you were to admire a building that has architectural elements in Golden Proportion and were asked to explain why you like it, you might say: "I don't know, I just do." A Pythagorean could be quite agreeable with your answer. A Pythagorean could also know that pressing you for more detail may not yield a rational answer – as when you may say, "Wow!"

Pythagoreans call irrational numbers *incommensurable* while the label 'irrational' appears about a hundred years ago. The irrational label is not incorrect but it defines the particular class of numbers by something they are not – rather than by something that they are. It may be worthwhile to call irrational numbers the *vibrating*, *hyperspatial*, or *infinite* (mantissa) numbers. Irrational numbers have many applications and other names such as *nodal*, *gap*, *geometric*, and even *foam*, *singing*, or *spiritual* numbers could be appropriate.

Proclus, a neo-Platonist, calls irrational numbers the *formless* numbers – an interesting characterization if one considers the irrationals as 'those that have no structure.' Comments by Proclus regarding irrationals are full of excitement and it is unthinkable he would consider the topic of irrational numbers the forbidden topic – then or in the past. Proclus' statement concerning irrational numbers is not only revealing but also mystical and full of implications. The overall meaning of his comments is that irrational numbers are dangerous in their own right – not unlike the content of Pandora's box, say. Proclus does not advocate keeping irrational numbers a secret but, instead, urges keeping the purpose and applications (needs) of irrationals a secret. Proclus has much respect for irrational numbers and, while formless, he treats irrationals as having a life of their own – and he speaks of irrationals as having needs. The irrationals, according to Proclus, can be so unpleasant their needs should remain hidden. Unlike Euclid, Proclus does not stop at defining what irrationals are not, and it is most insightful of him to use words such as *waves*, *image of life*, and *formless* in his comments about the irrationals:

> "It is told that those who first brought out the irrationals from concealment into the open perished in shipwrecks, to a man. For the unutterable and the formless must needs be concealed. And those who uncovered and touched this image of life were instantly destroyed and shall remain forever exposed to the play of the external waves."

Proclus statement certainly has a poetic quality about it – perhaps with a bit of chill running down the spine. In a reply, a more structured approach may be appropriate:

> "The dust bowl circles the sky black as it creams the water's edge, for the hex upon ether does not negotiate the first while attempting to scatter the second. I make machines, for I am not but a witness. I break machines, for I am not a slave to pretenders. I teach the truth to those who wish to work the stone without hammer or heat. I am the root of five and I have come and will be come."

Golden Ratio is one of the better-known irrational numbers that appears prolifically in nature and in the constructions of the pentagram, the pyramid, and in architecture. Golden Ratio consists of two lengths that are easily obtained through geometric means. If the shorter distance is normalized to 1, the ratio becomes 1.618... to 1. Golden Ratio expressed in its exact unreduced form is $(1+\sqrt{5})$ to 2. Golden Ratio is irrational even though it is constructed with integers only. The integers can be of any unit length and this accounts for myriads of nested and self-contained structures that can be made with the two lengths. Golden Ratio is at times called the Divine Proportion and has a number of interesting mathematical properties. The Great Pyramid at Giza has the shape of the Golden Proportion triangle that can be seen in the vertical cut at the midpoint of its base – the greater distance **a** is on the face going to the apex while its shorter distance **b** runs horizontally to the center of its base.

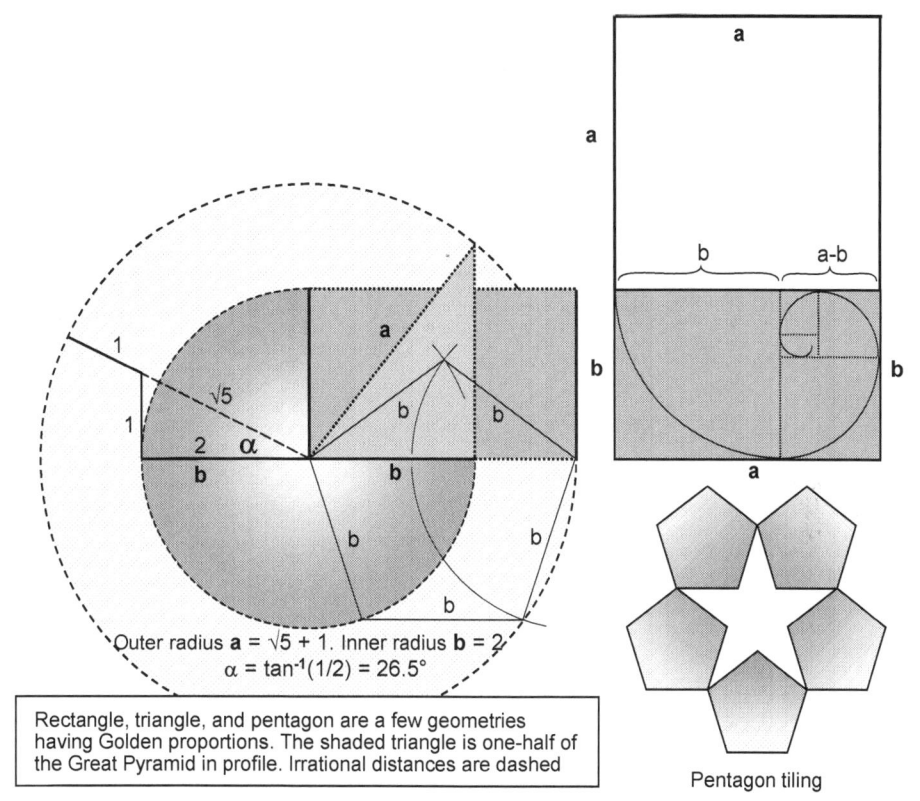

Quantum Pythagoreans

Outer radius **a** = √5 + 1. Inner radius **b** = 2
α = tan⁻¹(1/2) = 26.5°

Rectangle, triangle, and pentagon are a few geometries having Golden proportions. The shaded triangle is one-half of the Great Pyramid in profile. Irrational distances are dashed

Pentagon tiling

Golden ratio is in the *Golden Eye* as proportions of the outer and inner radii

Constructing the root of 5, adding 1, and relating the result to the length of 2 then directly constructs the Golden Proportion. Expressing the Golden Proportion in circular or "eye" format gives much freedom in the application of Golden Proportion. The association of Golden Proportion with patterns and geometries among diverse living things such as seashells, sunflowers, pineapples, and pinecones is quite remarkable. The angle α is also prominent in the construction of the Great Pyramid, as it is the ascending slope of the Grand Gallery.

Golden Proportion can well be expressed as a ratio, say, **a/b**. One of the interesting mathematical properties of the Golden Ratio is that $(a/b)^2 = (a/b) + 1$. Adding **1** to Golden Ratio is the same as squaring it and this is much help during construction. You may be tempted to reduce **a/b** into one number but keeping numbers **a** and **b** separate allows productive mathematical operations with **a** and **b**. For example, since one half of the Great Pyramid's base is **b** and its distance up its side is **a**, the

65

height of the pyramid can be computed with the Pythagorean theorem using **a** and **b** as separate numbers through the expression $\mathbf{a^2-b^2}$.[19] The central power of Golden Proportion is its ability to affect not only multiplication, division and (square) root through geometric means, but to concurrently execute particular computations among one and two-dimensional contexts. For example, the relation **(a/b) - 1 = b/a** also holds for Golden Ratio. By subtracting the integer value of length **1** from Golden Ratio you create a reciprocal of Golden Ratio, which could have either a spatial or scalar result. Simple constructions can then accomplish many complex mathematical operations.[20]

Golden Ratio **a/b** also holds for the expression **a/b = b/(a-b)** and this shows that the shorter distance **b** is in Golden Ratio proportion with distance **a-b**. Within Golden Proportion is an infinite set of Golden Proportions – each arising from *rotation* by fixed but particular number of degrees from previous proportions. Vast and diverse mathematical operations can be occurring concurrently and this is the mathematical "secret" of the Great and other pyramids. Once you become accustomed to applying the lengths **a** and **b** of the Golden Proportion in various ways, the thought of rationing and reducing **a/b** into one number will not have much value, and in fact may block you from making new discoveries. For example, once **a/b** is reduced, it is not possible to relate **a** to **b** with such questions as: "What does it mean when **a** and **b** relate as harmonic or geometric mean?" Presently, a fruitless search of old records was done in an attempt to uncover some symbol the ancient Egyptians would use for a reduced **a/b**, but what perhaps was discovered is that ancient Egyptians did not reduce the Golden Ratio into one number.

---

[19] As we enter the atom in the upcoming chapters the relationship $\mathbf{m^2-n^2}$ will become prominent in Johann Balmer's equation describing the atom's quantum numbers.

[20] Photons propagate or form a standing wave in one-dimensional geometry. In the upcoming chapters a review of atomic orbitals will end up with orbitals as three-dimensional due to electron spatial spread, while the atomic orbital will become two-dimensional at the moment of transition from one orbital to another.

Quantum Pythagoreans

Reducing **a/b** makes no sense if **b/a** and also **a-b** and **a²-b²** have useful applications.

Taking **a+b** and then multiplying it by 3/5 results in the expression (3+√5)·(3/5). The resulting straight (one-dimensional) distance is very close to π and approximates π to within 0.0015% or 15 parts per million. π is another well-known incommensurable number that, however, cannot be constructed as a distance between two points because π results from addition of an infinite quantity of contributing components. It is for this reason π is called the transcendental number; the infinite quantity of contributors does not allow π to be constructed as a distance between two points of length π using but a compass and straightedge in a finite number of moves. Inherently, π is required in curvilinear structures that close upon themselves by creating orbits and (atomic) orbitals. π is a numerical constant that converts the straight path having the even symmetry into a circular path that also includes the odd symmetry. The even symmetry is one-dimensional symmetry about the axis, while the odd symmetry is zero-dimensional symmetry about a point.

Here is one result that combines all of our present knowledge. If an object has a certain moving energy going in a straight line and begins to orbit another body, the energy must be conserved along a path that now includes π. Since the conversion of a straight path to a circular path is not exact, the idea of making a turn is not straightforward if the conservation of energy holds, for indeed it does. An object that begins to make a curve cannot conserve its moving energy exactly and so the excess energy must radiate in or out. Once the curving radius becomes constant the conversion is complete and energy inequality no longer exists. A motion along a spiral trajectory constantly changes the curvature, and unequal energy exists as long as the spiral path exists.

Numbers that relate to each other exactly through a finite multiple are commensurable. Ratios of any two integers always produce a finite rational number and all integers are, therefore, commensurable. A rational number is a number with naturally finite mantissa. Incommensurable numbers, on the other hand, do not have a common number that would change one number exactly into the other. There are two kinds of incommensurable numbers: transcendental and irrational numbers. Transcendental numbers express themselves in 2D while irrational numbers exist in 1D. Transcendentals, then, exist on a curve while irrationals exist on a straight distance between two points. Even though every irrational number has infinite mantissa, the Pythagorean theorem can construct irrationals in finite time. Transcendentals, however, are not constructible with the Pythagorean theorem.

Unlike π, the lengths that constitute the Golden Proportion are constructible. This means that at the end of the day (after a finite effort), there exist two pairs of points for each length and the distance between two points in each pair is the desired and exact quantity. For Golden Proportion the distance **a** is irrational while the distance **b** is rational. Golden Proportion geometries also allow infinite scaling and inclusion of ever-smaller components that continue to maintain Golden Proportion. Ever-larger construction is also possible because **(a+b)/a = a/b**, which allows ever growing construction while maintaining Golden Proportion. In the Appendix, Golden Proportion is applied in the nearly exact construction of π, the construction of the Alpha & Omega symbol, and in description of the principal purpose of the Great Pyramid.

The Pythagorean theorem relates distances of right-angled triangles' sides. However, there is one additional application that goes well beyond the simple and straightforward arithmetic of the theorem. The Pythagorean theorem describes the relationship between two-dimensional areas that exist on a plane and one-dimensional distance of a line. The square root is the operator that allows *area* size to be related to *line* distance. The square root is the critical operator that facilitates equivalence between the two-dimensional and one-dimensional entity. It is, for example, not obvious why a bar galaxy has the shape of a bar and why millions of solar systems happily line up in the shape of a stick. There just may arise considerations where a tractable solution can be had along one dimension but not in two dimensions. Unique relationships inherent in both the Golden Ratio and the Pythagorean theorem enable solutions to exist while tractable computing is subject to dimensional constraints. The simple interpretation of Pythagorean Tetractys as 'two dots are one-dimensional, three dots are two-dimensional…' begins to acquire a more practical meaning.

Pythagoras and his fellow Pythagoreans discovered that a single string pinched in ratios of certain whole numbers resulted in clearly vibrating musical tones. A musical instrument such as a flute, moreover,

Quantum Pythagoreans

adds harmonics or overtones to the tone being played. Harmonics are additions of half-integer multiples of the tone being played and different multiples give different instruments their uniqueness, also called timbre. However, when a string is pinched with proportions of incommensurable (irrational) numbers, vibrations contain many frequencies that are not simple multiples of some average or dominant tone. Certain vocal sounds such as 'om' and 'aum' also have rich content of frequencies. The playing of some irrational numbers gives fullness to tones but other irrational numbers can create unpleasant and even disturbing sounds. An irrational number that is not reduced into a finite real number inherently contains a superposition of an infinite quantity of components in the form of ever-higher and ever different frequencies. A reduced number, on the other hand, becomes the real number that contains just one value and acquires fixed and finite magnitude. The *unreduced* irrational number retains not only all of its frequencies but also retains the spatial distances and angles that are inherent in an irrational number's creation. The irrational number is reduced in a computer but the irrational number is not reduced if it is geometrized as non-solid diagonal, or if it is architected, played, sung, visualized, watched, or admired. Some irrational numbers such as the Golden Ratio give pleasant impressions and the variety of life forms also support the Pythagorean notion that the pentalpha (pentagram) and pentacle are healthful and life promoting. Some other irrational numbers, however, sound shrill and harsh, and may even be detrimental to ordered systems. It is quite likely that differentiation of irrational sounds into pleasant and unpleasant sounds is absolute – that is, our upbringing has no influence on how we interpret irrational sounds. Some sources describe Pythagoreans as understanding both the pleasant and unpleasant nature of sounds produced through incommensurable proportions. Pleasant sounds also have healing properties while unpleasant sounds have detrimental properties.

There is much to say about incommensurable numbers that borders on mystical. It is possible to construct the length of two exactly. It is also possible to construct the exact right angle line of length one that results in two right-angled arms of exact lengths. But the diagonal distance that now exists from one end to another cannot be constructed exactly. The diagonal is constructible because there exist two points defining its beginning and its end. However, the *length* of the diagonal is not constructible exactly because the length is an incommensurable number. There now exists a gap because the right-angled construction from the one point to another is exact and is constructible exactly, but in the straight line construction of the diagonal the length is never exact, no matter how small the gap gets.

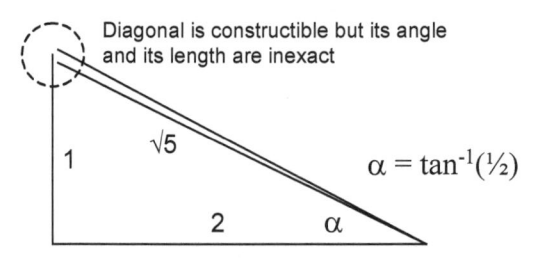

Diagonal is constructible but its angle and its length are inexact

1, √5, 2, α, $\alpha = \tan^{-1}(½)$

Gap where the exact meets the infinite

From the perspective of real numbers the gap can be ignored – the straight interconnection between the two end points completes the construction. However, the closure of the gap through solid means also reduces the incommensurable number into a real number because the solid closure acquires exact distance.[21] By allowing the gap, though, all frequencies associated with the root of five now have the opportunity to express themselves. The distance of the square root of five now becomes a family of infinite number of frequencies.

The straight incommensurable (irrational) distance is the *centerline* of a family of standing waves. The angle of the centerline can be computed from trigonometry. The mystical part is that the centerline exists because the two points of the line exist – yet the length of the centerline is not constructible exactly. The fact that the distance is not exact allows the expression of the infinity that is inherent in every incommensurable number. The centerline may be bounded and there are two constructible exact lengths: one length is greater and one length is shorter, but the length and the angle of the centerline itself cannot be

---

[21] Some writers see the truncation of irrational distance as the proof of the inexactness of the world and lament the world's finiteness as well as the loss of data. It is said Alexander The Great shed tears because there were no more worlds left to conquer.

Quantum Pythagoreans

constructed exactly. But if the incommensurable distance is to be realized, it must exist as open distance between two points.

In the following illustration, the actual implementation of the √5 number is taken from the Great Pyramid. The edge of infinity is the Great Step at the top of the Grand Gallery, while the centerline of the ascending slope is matching the 26.5.. degrees issuing from **tan⁻¹(½)**.

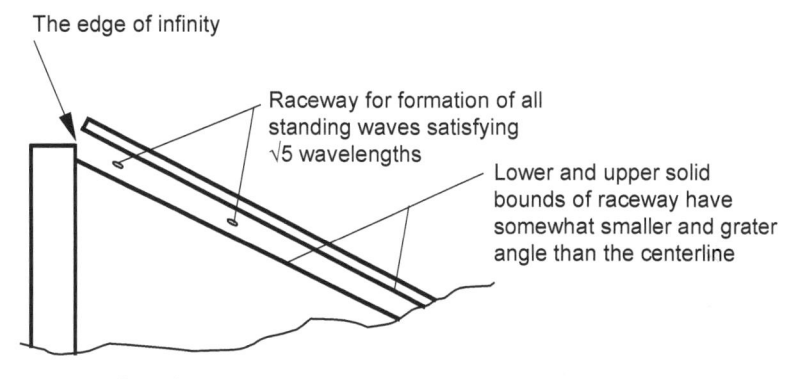

The edge of infinity

Raceway for formation of all standing waves satisfying √5 wavelengths

Lower and upper solid bounds of raceway have somewhat smaller and grater angle than the centerline

Standing wave centerline guide

The Pythagorean central tenet is: *All is number*. The meaning of this phrase becomes apparent when the number actually becomes – or comes alive. The number may be written on a piece of paper but that is but a representation of such number. Every number can be constructed to become useful and create things that can have a life of their own. Numbers, then, come into being and that is the meaning of *'All is number.'* Presently, the incommensurable numbers are becoming...

The infinite expression of waves exists in the interconnection through the gap and along the centerline, which can be visualized as an open, non-solid, standing wave guide where each and every wave from among the infinity of waves can find its own beginning and ending node. Each and every wave can find its own fit and thus its own existence. While the quantity of wavelengths is infinite and all waves exist in infinite superposition, there are but those particular wavelengths present that satisfy the construct of the root of five.[22] One node shown in the illustration above is labeled the *edge of infinity* and at this point all frequencies are anchored – that is, the nodes or Newton's fits of all frequencies are to be found at this edge. Solid bounds that define the width of the gap also act as filtering bounds that do not allow other frequencies to form. Many accommodations for infinite wave formations exist inside the Great Pyramid. Some triangles such as the right angle triangle with lengths of 3, 4, and 5 have all sides of exact length and then the presence or absence of a gap is not germane. Right-angled triangles that have the exact length on all three sides cannot form a superposition of infinite wavelengths and are rare in the Great Pyramid.[23]

Twenty four hundred years ago, Euclid compiled and organized the knowledge of geometry that came from many sources. While it continues to be true that there is but one line that is constructible between two points, the length – that is, the magnitude – of such a line is not always known exactly and can be indeterminate. Up to the present time the drawing of a solid line connecting two points is not thought to diminish Euclid's work. It is always easy to presume that, even if the distance is not exact, it is close enough or can be made more exact if needed. But that is not the idea. The incommensurable distance, if not truncated through solid means, offers significant opportunities for working with waves. A new axiom that is added to Euclid's axioms now becomes important because it creates a gateway for new applications. Euclid understood incommensurable numbers in terms of their definition,

---

[22] In Aztec and Inca traditions, our world is the world of the fifth sun. In the present context, the root of five is purely a mathematical construct, although the family of frequencies associated with the root of five can well be associated with a particular life form – that of our own. It is likely that the protein creation by way of three-dimensional folding is possible in an environment that is saturated by the frequency content consisting of the family of the root-of-five frequencies.

[23] In the upcoming sections the eigenstate is introduced as a solution to computations that consists of but real components. Right angle triangles with diagonals of exact length can be seen in the context of arriving at real-only solutions. In the atom, eigenstates are but temporary states that quantify real energy for the purpose of the conservation of energy at transformation.

which meant that the irrational number couldn't be constructed through division that is rationing of two whole numbers – and he presents a much-copied proof to that end. It is likely that Euclid, unlike Pythagoreans, did not think of incommensurable numbers as having non-repeating and therefore infinite mantissa, which, in turn, could have raised the question of exactness. Pythagoreans, for their part, also called incommensurable numbers the *unspeakables* and it is likely they understood the incommensurable numbers as being infinitely long and, in fact, indescribable. Moreover, looking at the Pythagorean culture of secrecy from the outside, it is easy to presume that 'unspeakable' is synonymous with 'nondisclosure.' To a Pythagorean, however, something that is unspeakable also means that 'words are not enough' to describe it, and that nonverbal means are better suited for working with incommensurable numbers.

The addition to Euclid's axioms is as follows: There exists but one line of finite magnitude between two points spanning rational distance; and there exist infinite superposition of waves between two node points spanning incommensurable (irrational) distance.

Pythagorean mathematics has structural roots that allow the concept of force and force itself to arise. The force happens through mathematical and spatial relationships that are separate from, say, the engineering concept of the electrical generator that needs to physically rotate to produce electrical energy. Non-mechanical energy transformations then also play a pivotal role. Transformation needs a *from-to* context, which Pythagoreans equated with even and odd number groups as well as with incommensurable and real number groups. Geometries play the central role because certain information – or knowledge – can be acquired only through certain geometries. Shapes and forms then also play a role in acquiring or creating knowledge that cannot be obtained or created otherwise.

A Pythagorean mathematician does not just describe the behavior of nature. He or she also strives to create what nature creates. A real thing is something that was and can be made with numbers through a mathematical and geometric relation. Water, for example, is a computable structure that is constructed through computational means. A Pythagorean mathematician can also arrive at a mechanism that allows remote control of the chain reaction in unstable matter. A Pythagorean mathematician can do all these things because matter and its stability are fundamentally computational and the pursuit of the fundamental truth allows its unveiling.

Numbers Express Variables

A variable is a changeable attribute of some entity. When entities interact with each other, their variables influence and change each other. If variables influence each other in computable fashion then the change between variables can be calculated and the values that the variables acquire become predictable. Variables are computable when arithmetic exists that relates variables to each other. The computable relationship is called a *function*. The function can be charted to show one variable versus the other and provide a better picture of the influences between the two variables. It is just such influences that can reveal some of the most fundamental relationships in nature.

Moving energy E equals $½m·v^2$. The two variables of mass m and velocity v can be charted where one or both variables may be changing while the energy is expressed by the function $E = ½m·v^2$. Leibniz's work established the formulation and the conservation of moving energy. From there Leibniz was the first to formulate the conservation of momentum that explicitly included the *direction* of the moving body as a parameter that, in addition to energy, is being conserved. Leibniz also worked with the analytical framework of 'coordinates' and 'axes of coordinates' of René Descartes. Leibniz thought so highly of his analytical methods he started to describe metaphysical concepts with his methods. Newton, while being a fertile writer on metaphysical topics himself, did not feel comfortable enough to publish any of his metaphysical or alchemical manuscripts.

Dependence And Independence of Variables

The simplest function has two variables, usually labeled x and y. Current convention calls variable x the independent variable while variable y is called the dependent variable. Variables on the right hand side of the equation are presently called independent because the mathematician can technically choose any desired value for the right-sided variable. The calculation then yields the "dependent" variable on the left hand side of the equation because the result depends on the arithmetic operations performed on the independent variables. The present convention of dependence and independence is too simple to be of much use, so we will replace the present definition with a new treatment. Moreover, the rules of algebra do not differentiate between the so-called independent and dependent variables. In algebra, any variable can be placed on the left hand side of the equation and, algebraically speaking, calling any variable dependent or independent is arbitrary.[24] Algebra does not capture the independence and dependence of a variable even though it is apparent that classifying variables as independent and dependent is desirable in some way. Another way of looking at independence and dependence is through the recognition that there are variables that lead and variables that follow.

The independent and dependent labeling of variables originated in classical physics where direction, distance and time, for example, are always considered independent. Classically speaking, direction, distance and time can be "anything they want" and the presumption then exists that variables such as these have no restrictions in terms of values.

The new definition of dependence and independence of variables comes from the mutual values variables have at some region of their relationship. There is, for example, a likelihood that the variable of spatial distance can be independent at times but becomes a dependent variable at other times. Spatial distance can at times be anything and thus *independent*, while at other times the distance a body may negotiate becomes subordinated or *dependent* or restricted or influenced by other variables. A particular variable may acquire a value that is a result of restrictions imposed by other variables, and such restrictions then determine the particular variable's degree of dependence. Lack of restrictions placed on a particular variable then results in independence.

---

[24] Basic rules of algebra were described by Al Khowarizmi and translated into Latin in the 12th Century.

●●○○○○○ **Star Numbers and Operators**

Dependence and independence of variables is between variables themselves, and the position, whether on the left or on the right hand side of the equation, is irrelevant. A particular variable that is dependent in one context can become independent in another context. A particular variable can be simultaneously dependent with respect to one variable while being independent with respect to another. Also, a variable that is dependent with respect to a variable can become independent with respect to the same variable when circumstances change.

The *function* is the mathematical description of the strength of variables. The function that cleanly shows the dependence and independence of variables is the hyperbolic function, illustrated below. It is said that the use of the word hyperbole in mathematics and in English evolved separately but some say they have the same meaning. 'This book weighs a ton,' is a dictionary example of a hyperbole where the book feels so heavy it makes no difference if the book weighs two pounds or two thousand pounds – and its actual weight is irrelevant. When graphed, the weight of the book is beyond a point called the Extreme point. Tracing the curve in bold, from the Extreme point outward, the weight of the book can increase without bound but the book still "weighs a ton."

If the relationship is along the mid section of the curve – shown as a thin line – then either variable can be independent or dependent while the dependence and independence may be reversible. In the sentence 'Plan your work,' **plan** is the independent variable, for this can be any plan, whereas **work** is the dependent variable because all **work** is for the purpose of planning and all work is being planned. The sentence could be reversed as 'Work your plan' where **work** becomes the independent variable and all plans are now subject to being worked. Some hyperbolic reversals do not hold and lose meaning but many hyperbolic reversals continue to have real or figurative meaning. Some reversals, such as 'Quantity sold depends on price' and 'Price depends on quantity sold' have valid but not identical meanings. Similarly, 'Take the accused to the justice' is reversible but results in different meaning. Surprisingly, dictionaries or English textbooks do not define or treat the independent-dependent reversal at all and one can only surmise that the accepted concept of the hyperbolic reversal exists as but a collection of numerous expressions or sayings – as if the hyperbolic reversal of 'Sword is mightier than pen' does not need any more ink. The independent-dependent reversal will be called the hyperbolic reversibility or just reversibility.

'Quantity Sold' can be readily put on one axis while 'Price' can be placed on the other axis. Non-numerical relationships may not be as easy to place on the axes yet some relationships will continue to have meaning even as the hyperbolic reversal flips the independent and dependent relationship between the two variables. Once a variable passes the Extreme point, however, then for all practical purposes such a variable becomes independent of the other variable – it is past the Extreme point that the English dictionary defines as a hyperbole. The statement "This book weighs a ton" is an example of a relationship at the extreme portion of the hyperbole where it is no longer reversible. For the purpose of organization and self-organization, however, all relationships will exist between the extremes that are on the thin hyperbolic curve.[25]

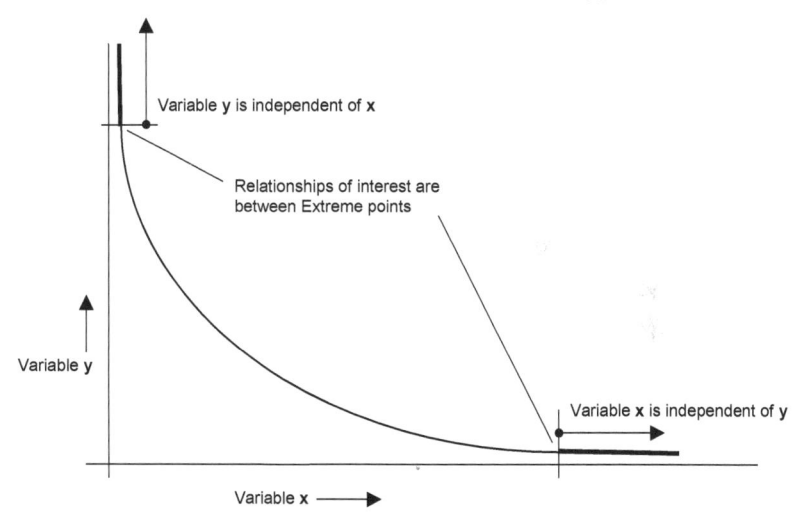

Hyperbolic function: Variables **x** and **y** in hyperbolic relationship

---

[25] A point can be made regarding the well-known Buddhist path of the Middle Way and whether there is similarity with the context where variables are being worked "between the extremes." The critical and desired qualifier is the *reversibility* potential of the dependence and independence of variables, which can happen only in the middle and between the extremes.

As variable **x** increases along the horizontal axis there is a point beyond which variable **x** can increase dramatically but **y** will not change much. Beyond this extreme point, variable **x** can be "anything it wants" and thus variable **x** is *practically* independent. In the example of the English language hyperbole 'This book weighs a ton,' the weight of the book (a ton) has passed the extreme point: The weight of the book is overriding all other considerations and the parameter 'weight' is now almost completely independent of other variable(s). Mathematically speaking, beyond the extreme point on the horizontal axis the variable **x** is no longer a function of **y**. The independence, however, reverses when variable **y** increases beyond some point with respect to **x**, since variable **y** can now vary and this variability has no significant influence on **x**, and now variable **y** is practically independent of **x**. Overall, as the drawn function moves toward horizontal or toward vertical the independence of one variable versus the other variable is easy to visualize. Hyperbolic function is useful in classical physics and in quantum mechanics as well. Hyperbolic relationships come up often when the mechanics of self-organization of matter is discussed because hyperbolic relationships transcend the apparent separation of the macro world of cosmic mechanics and the micro world of atomic physics.

Because the dependence and independence exists between variables, no single variable is inherently independent. The fact that no variable, physical or otherwise, can become totally independent is very fundamental and becomes the first principle. No variable is intrinsically independent and no variable can make a claim to be "anything it wants" at all times and under all circumstances. The state of dependence and independence exists between at least two variables and total independence would be about some variable that does not relate to any variable and such variable could not relate anything and thus relate to nothing. The conclusion from this is that no entity exists that could not be detected under some circumstances. Yet another way of stating it is that there exists no entity that could not be influenced. The first principle can have many different kinds of esoteric names such as the principle of all-connectedness or wholeness, the principle of all there is, the principle of not one, or the principle of plurality. The first principle has a touch of Tao, for there is no second principle – only one principle of not one.

Philosophically, the first principle is symbolized in the West as the *locus* in the geometric tradition. In the East it is the lotus blossom.

Time is one variable that is always a dependent variable. Time can never become the independent variable because time is a derivative that always issues from variables that have periodic property.[26] Lightspeed has a periodic property because light periodically – that is constantly and repeatedly – traverses a particular unit of distance. When time is derived from a variable that happens to be independent then it may appear that time is also independent. However, when such independent variable becomes dependent, time will also become dependent; time cannot become independent with respect to a dependent variable. Time's relationship with other variables is always that of a derivative and that also means that the relationship between a variable and its derived time is not reversible. Because time is a derivative from an independent variable then *time overlay* is a good working term for time. Time can then be mathematically treated as an independent variable as long as the variable from which time is derived remains independent. Given that so much has been said and written about time, a separate chapter will deal with it.

The concept of independent and dependent variable goes beyond the assertion that the independent variable leads and the dependent variable follows. The definition of the independent variable is that it is *constructible and holding*, while the dependent variable issues from the independent variable. In an example, energy is applied such that matter and antimatter appear, but matter and antimatter arrangement does not hold and the existence of leading and following variables does not even arise. In another example, a triangle has the sum of its internal angles always equaling 180 degrees. Euclidian geometry proves this in a straightforward fashion. However, when the triangle is on a curved

---

[26] In Buddhist vernacular, the derivative – or a variable being derived – is a variable that exists *by convention*, or as having conventional existence. Some Buddhists may call the independent variable the absolute variable. This also works, particularly in the context of measurement. Yet using the word 'absolute variable' carries some implication of invariance, which may create unnecessary confusion at transformation.

surface – such as may be the case under point symmetry – the sum of the three internal angles in a triangle exceeds 180 degrees. Presently, much discussion is on whether the flat or curved geometry is the right one. Both geometries, however, have their own unique applications: Euclidian geometry holds for straight, one or more-dimensional construction of the standing or propagating wave while the curved geometry of Georg Riemann holds for atomic geometry with closed and finite orbitals of curving waves. But even if we all were to agree that both hold because Euclidian geometry holds for straight topology while Riemannian geometry holds for curved-and-closed topology, the essential question to resolve is: are both the flat and the curved topologies individually constructible? That is, can both topologies be constructed and remain valid in their own right? Better still, can it be shown that the system under consideration exists and remains in one topology and not in the other? Is, then, a given topology enforceable? Light, being a virtual entity that is not and cannot be subject to gravitation or other electromagnetic field influences, will always enable the enforcement of the straight Euclidian geometry. Any deviation from the straight path of light can be identified and the flatness of Euclidian geometry enforced. A laser, for example, enforces straight lines and flat areas – and triangulation over any distance will result in internal angles of a triangle to sum up to 180 degrees. The differentiation and individual enforcement of one, two, and three-dimensional values is symbolized by Pythagorean Tetractys. In the upcoming chapters the pyramid will be introduced in the context of transforming the standing waves or straight-moving waves into curved and closed (localized) geometries of the atomic orbital. Yet another way of seeing the separation and enforcement of dimensional independence is through transcendental variables. It takes transcendental variables to operate in Riemann geometry and the actual construction – that is, the active creation – of transcendental variables is required to transform between Euclidian geometry that is straight, and Riemann[27] geometry that is curved.

---

[27] Riemann family is from the Slavic region of Germany and some claim Riemann is a Germanized name of Říman – Slavic for Roman – a citizen of Rome.

Quantum Pythagoreans

'Squaring of the Circle' is an exercise in proving or disproving that a particular curving trajectory such as the circumference of a circle can be geometrically expressed as linear distance. While many have tried, recreationally and professionally, the construction from a curved topology into a straight topology of exactly the same length is not possible. This result, then, points to a hard separation between a one-dimensional and two-dimensional topology and, moreover, it shows that the linearly expressed energy of linear motion does not map exactly into the curved energy of the orbiting motion. Because the energy is conserved, there arises an energy imbalance when a motion of a real object changes between linear and curving paths or vice versa. It should come as no surprise, then, that straight and curved topologies exist in independent as well as absolute contexts, and are not subject to convention.

When the topic of cosmic topologies comes up, there will be accounting for the simple fact that although a variable is independent by wielding the largest influence, the dependent variable's influence never becomes zero. This can be seen as a restatement of the first principle where no variable can ever exist that could be so independent that it would not relate to anything else. There is a certain spiritual quality to this because an assertion can be made that no entity can ever be split in such a way as to make any of the resulting components completely independent of each other.

In economics, as the product manager increases the price of the product, the quantity sold decreases and approaches zero as the price continues to go up. In reverse, lower prices spur an increase in sales, and as the price approaches zero, the quantity sold goes up and up. At one end the price is so high that even a large movement in price fails to spur sales. At the other end the quantity sold is so high that sales variations – with price already very low – have little to do with price. The relationship of price versus sales is along a hyperbole, and marketing people like to talk about points on the hyperbole as having different elasticity. A starting price that can increase without significant limiting or offsetting effects becomes the independent variable, producing positive and good elasticity – as far as marketing people are concerned.

When two bodies gravitationally or magnetically interact, their distance and their force of attraction relate to each other along a hyperbole. On the one end, bodies are far apart and the distance can change a lot without any appreciable change in the force between bodies. On the other end, however, force dominates and can change a lot without appreciable change in the distance between bodies. In this case it is a symmetrical hyperbole where the independent property of distance and the independent property of force can reverse. In this case either the distance or the force each have a region where each can claim to be "anything it wants." Either variable can have its own independence where one variable is not subordinated to the other variable. 'The tail wags the dog' is a hyperbolic reversal where, taken literally, the dependence and independence cannot be reversed.

Many hyperbolic relationships add up and form our perception of reality. If some things happen in a particular way over and over in a similar environment, the brain forms hyperbolic functions that tell us which things are dependent and therefore subordinated, and which things are independent and therefore dominant. To see a tree branch being sawed off but only the tree trunk falling down happens in cartoon land. The tail may not wag the dog but a person succeeding against great odds does just that.

Present day mathematics generally does not deal with the dependence and independence of variables. The statistic of correlation does not provide a clue to the dependent-independent aspect of variables because correlation is but a measure of variables' concurrent direction. Correlation is not concerned with which variable leads and which variable diminishes or if such a relationship can reverse. An increase in outdoor temperature correlates positively with the consumption of ice cream but large or small consumption of ice cream will have no effect on outdoor temperature. Yet, in statistical correlation the variable **X** correlates with **Y** to the very same measure that the variable **Y** correlates to **X**. Correlation cannot claim which variable affects which. It is, however, obvious that the outdoor temperature affects the consumption of ice cream and not the other way around.

The consumption of beverages such as coffee can be correlated to a number of ills but there is no indication what variable leads and what variable follows. Does the consumption of coffee lead to a particular ailment or does a particular ailment stimulate the consumption of coffee? Or is it, perhaps, that the consumption of coffee diminishes the severity of a particular ailment? Statistics of correlation cannot capture the directional relationship between two variables. On the other hand, the independent (leading) and dependent (following) qualities of a variable do capture the directionality of a relationship and advance the understanding of relationships beyond symmetrical correlation.

Causality And Priority

On both the abstract and practical levels, philosophers as well as engineers and lawyers are always greatly interested in causality. Cause and effect is a sequential process that is repeatable and, therefore, predictable. There is much value in causality because it lets us construct sentences using 'because.' The *if-then* construct indicates causality. Causality springs from a particular reason and points to a purpose of some action. Causality has a flow and one thing (reason) leads to another (purpose). Causality needs and does have the benefit of precedence that reveals what started the causal process. Causality has legal ties to the concept of negligence and fault because repeatability and predictability opens the issue of avoidance. In a valid *if-then* construct the causality is at its strongest when the absence of the causal supposition *if* altogether invalidates the outcome *then*. Current computer technology cannot function without the *if-then* construct because the computer needs to recognize a particular condition *if* to make a particular decision and take one or more specified actions. For example, 'if the temperature exceeds 150 degrees, (then) sound the alarm and close doors' could well be a group of computer instructions that checks the supposition *if*, and if true, another group of instructions is executed. No other condition would cause the alarm to sound and the instance of a temperature exceeding 150 degrees is a unique and true causal event.

Aristotle established four components that make up causality: matter, form, moving cause, and final cause. The present day computer fits the definition of Aristotelian causality quite well: (**1**) matter – the silicon; (**2**) form – Boolean (after George Boole) logic gates for hardware, which also decode address and microcode of software; (**3**) moving cause – the trigger of the satisfaction of the *if* construct; and (**4**) final cause – the execution of specific instructions once the *if* construct is satisfied. It appears that by restricting the first causal component to matter, Aristotle delegates the causality to hold among real things only – a nice insight.

It is fashionable to say that in the quantum mechanical environment the observer and the observed become entangled in some such way that causality weakens to the point of possibly disappearing altogether. Quantum mechanics, however, is about relationships between variables. In the chapter on photonic branching it will become apparent that interactions in the quantum mechanical environment do follow certain rules. Even on the quantum mechanical level the relationships between photons and electrons, for example, are well evolved and once the relationships are confirmed experimentally, the outcome can be repeatably forecast. Quantum mechanical relationships are logical interactions that can be visualized as influences, where some influences prevail over other influences. Logical interactions among influences usually do not manifest in physical action. Causality, on the other hand, always manifests as physical or real action. When logical interactions among influences do manifest, the transformation takes place; in the quantum mechanical environment such an event is called *reduction*. There will be more to say on reduction in the upcoming chapters.

If the hyperbolic relationship exists between two variables the independent range of one variable compared with the dependent range of another variable yields characteristics with significant practical applications. In physics the independent variable is also called a variable having *priority*. A variable with priority prevails over another variable in their mutual encounter. One variable may be found dominant in its interaction with many other variables so qualitative as well as quantitative one-to-many relationship can then be established. The independence, leadership, dominance, priority or similar designation of a relationship that one variable has with respect to another then also fits well with the ancient and honorable notion of causality – especially if the relationships are repeatable and particularly if such relationships result in physical action. In the absence of manifestation of physical action, quantum mechanics continues to be relevant through the computational change in parameters' value – that is, the outcome of future events has been changed even though there is presently no overt manifestation of such change. Change through computational means differs from

Aristotelian *potentia*, which is the innate tendency or unfulfilled propensity such as 'bubble has potentia to form a sphere' or 'salt has propensity to absorb water vapor.' In quantum mechanics the changes in the values of a variable is due to a priority that may appear similar to potentia, but the values can increase or decrease and can only be changed computationally: Electron wave shape is recomputed through interactions with matter or electromagnetic fields, and then the *probability* of the electron reappearance at a particular location has been increased or decreased – now and into the future. Aristotle can take heart in Erwin Schrödinger's equation, for it will be shown that the electron has potentia to evolve in time, which happens when the electron acquires one or more degrees of independence.

In classical and in quantum mechanical physics the interactions among variables are repeatable within a certain range. Leadership of one variable with respect to another can reverse, however, which is something the classical Aristotelian causality does not readily recognize. Aristotelian compatriots, however, could agree with a contemporary study concluding that wine is healthy and reduces diseases but in large quantity actually promotes diseases. Further, a variable may have priority over another variable while losing its priority to a third variable. The context-dependent and range-dependent change in priority is inherent in relationships. Causality, however, applies to a particular objective action and so the context and ranges are fixed to assure the repeatability of the causal process.

In a free economy the economic variables are generally not fixed. Leadership may be quantified in some context but later on the value of leadership may change even if the context remains the same. Interest rates and the rate of inflation have strong leadership measures but after some period recalculation will likely result in another value. Similarly, changing the length of the historical period from three to six months changes the context of analysis and results in different leadership values.

A photon of light interacts with a free electron in some way. Certain modeling would have light bouncing or deflecting the electron from its path if the model uses the presumed photon's real momentum and presumed solid particle electron to determine causality and thus the outcome. A free moving electron, however, transforms its momentum into vibration and now the independence-dependence relationship may change so much that a new experiment may be called upon to find the head and the tail again. If the interaction between photon and electron does not result in physical and observable change, the interactions are that of priority instead of causality.

At The Gate To The Cave Of The Shadows

> *"...and he can read the defense, too!"*
> Plato as a sports announcer

The cave of the shadows is a metaphor attributed to Plato. Plato was a Pythagorean who established an academy in Athens lasting 900 years – an unsurpassed achievement. Plato said and wrote about many things, including the application of numerical relations to the running of the state. What lives on is Plato's metaphor that espouses the indirect way of looking at the world. Through the story of a cave, only reflections in the form of shadows are available to the participants. In the East, the cave of the shadows metaphor is similar to the Buddhist finger-at-the-moon figure of speech that shows how difficult, if not impossible, it is to reduce into words complex symbols such as the moon. Presently, the best example of the indirect method is by describing the force. It is not possible to describe what force is. The only practical way of dealing with force is to describe how force manifests and what it does.

In addition to seeing and describing all possible things around us, many events happen as a result of influences that are difficult if not impossible to pin down. Causal links may become altogether absent and there is no solid ground on which to build our reasoning. Worse, links may be all around but they are so weak they may appear irrelevant and may in fact be irrelevant. Such links exist in forms of hints and suggestions rather than clearly stated causal links. Worse yet, good causal links may exist but following such links may take so much time the process becomes intractable. There is, then, but one opening and that is to assemble the case circumstantially. That is how we are going to enter the cave of quantum mechanical shadows, for quantum mechanics is about things virtual, things intangible, things altogether frustrating to all who like to think that everything is connected to everything in a predictable and measurable way.

Dual Slit

Truth is something that works in more ways than one

The dual slit experiment is about two slits that are narrow and closely spaced. A lone electron is directed at the slits and as the electron moves it carries de Broglie vibration. It is not difficult to accelerate electrons to the same desired speed so all electrons have identical wavelength. With its wave characteristic the electron goes through both slits simultaneously and, landing on a screen on the other side, makes a dot on the screen. The second and the other electrons that follow, however, do not always land on the same spot and after a while a pattern is discerned called interference. While this experiment is a factual experiment that was duplicated with other small particles in addition to electrons, the underlying mechanism has eluded explanations. In the explanation that follows, it is easier to discuss photons of light first because the interference pattern forms with photons as well, and with photons there is no need to be concerned with the parameter of mass.

The interference pattern electrons make is qualitatively the same pattern as made by coherent light, which was first described by Thomas Young in the early 1800s. Coherent light is presently available directly from a laser. All laser photons have the same wavelength and are thus coherent even though individual photons do not come out of the laser synchronized in clusters or at the same fixed separating distance. Coherence helps in calculating the results because but one frequency needs to be considered. From a laser light source, then, a light beam is sent through two vertical slits and the interference pattern forms at the screen behind the slits.

● ● ○ ○ ○ ○ ○   **Star Numbers and Operators**

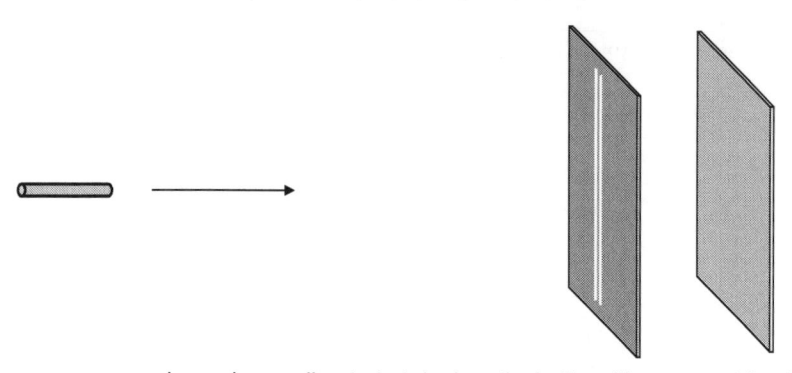

Laser beam directed at dual vertical slits with screen at back

For light, the interference pattern forms when two waves, one from each slit, are added to each other for a brighter (more intense) shade or subtracted from each other for a darker shade. At points along the screen, one of the wave's traveling distances differs from the other because the separating distance of the slits gives each wave a unique starting point.

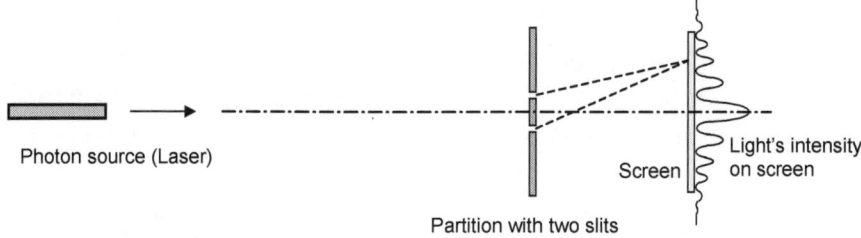

Photon source (Laser)   Partition with two slits   Screen   Light's intensity on screen

Two paths from slits to screen have unequal length, Top View

Two light wave crests come out of the two slits *simultaneously* but arrive at different parts of the screen along paths of different lengths. If the two waves meet at the screen at their maximum then that is where the point of highest light intensity will be. Similarly, two waves meeting at the screen with their minimums results in the darkest spot. Coming out of the slits simultaneously can be also described as being in-phase, phase-locked, or as being synchronous.

The simultaneous exit is crucial because if light crests were not leaving the slits in-phase then the interference pattern would not form. If photons were to pass through either one of the slits as individual photons at their individual times the interference pattern cannot form but instead a two-peak pattern would form – one peak for each slit. The first conclusion is that each and every photon must part because that is the only way for simultaneous departure to happen, which, in turn, is a necessary condition for interference. Indeed, if only one photon at a time is sent toward the slits the interference pattern continues to build. After each photon passes through both slits it makes a single point of light on the screen and over time the pattern emerges. Because every single photon parts and goes through both slits simultaneously the resulting interference is happening with each and every photon and that is why the interference pattern is also called self-interference. Knowing the separation distance between slits and the photon's wavelength, the interference pattern can be computed and the photon's frequency was found in agreement with the actual field data.[28] While the most intense light lies on the axis of the photon source, light's intensity (photon quantity) decreases away from the axis because greater deviation from the axis is less likely to happen. It will become apparent that each end of the parted photon acquires some degree of independence, and consequent movement away from the axis results from geometries the photon is passing through. The geometries of various shapes and sizes determine the outcome of photonic interactions. The next chapter on photonic branching and spreading will further examine the movement away from the axis.

Both photons of light and electrons produce the same interference pattern when presented with a dual slit. The sameness of the pattern drives the initial assertion that, since the photon passes through the slits simultaneously, the electron passes through the slits simultaneously as well. Here is the case so far: While it is quite apparent that each photon parts and exits both slits simultaneously, it is not certain that the electron goes through both slits as one entity. But Davisson and Germer confirmed electron spreading because the electron must interact with many thousands of atoms on a crystal surface in a simultaneous manner if an electron is to rebound as a wave. It is, therefore, certain that

---

[28] If the light source is not coherent and different wavelengths of light are sent at the slits the interference pattern still forms but the overall pattern loses sharpness because the overlapping interference patterns each correspond to different wavelengths.

the electron spreads and arrives at the entrance to both slits as a spread-out electron. In addition, the electron can produce the interference pattern only when the electron leaves both slits simultaneously. If a single electron is arriving at both slits simultaneously and departing both slits simultaneously then the only unknown is how the electron makes it through the thickness of the slits. The Davisson and Germer electron spreading indicates that the electron spreads in 3D and becomes a 3D entity. By the act of spreading not only in 2D but in all three spatial dimensions, the electron can and does pass through both slits while maintaining one spatial volume at all times and, therefore, the electron does not split up into separate volumes as it is going through both slits. De Broglie started on this road when he mathematically equated a real moving electron with a vibrating electron. Schrödinger, in turn, mathematically described the Davisson and Germer electron wave behavior, because the wave nature of the electron can express itself only when the electron becomes nonlocal. When the electron velocity is known, the electron's wavelength is also known. The interference pattern is then computed and the resulting shape predicted and verified.

For electrons the interference pattern is visible as stripes of various shades of gray. While one electron realizes (reappears, materializes, "lands") in a particular position on the screen, the next electron realizes in another position and over time the interference pattern becomes apparent even though only one electron at a time simultaneously passes through the slits and one electron at a time becomes visible on the screen. A particular stripe on the screen is a lighter shade of gray as electrons accumulate there in greater numbers. Since electrons collect at particular regions of the screen with different probabilities, many electrons need to go through the slits before a pattern emerges. A row of electron counters can be used instead of a screen and a cumulative electron count kept for each counter. With counters, the interference pattern will look similar to the graph below.[29]

---

[29] The graph is computationally produced such that each free electron is simulated as a string of random numbers that is processed in a two-stage engine. Different "electrons" produce different operations count when processed by the engine's procedure.

Quantum Pythagoreans

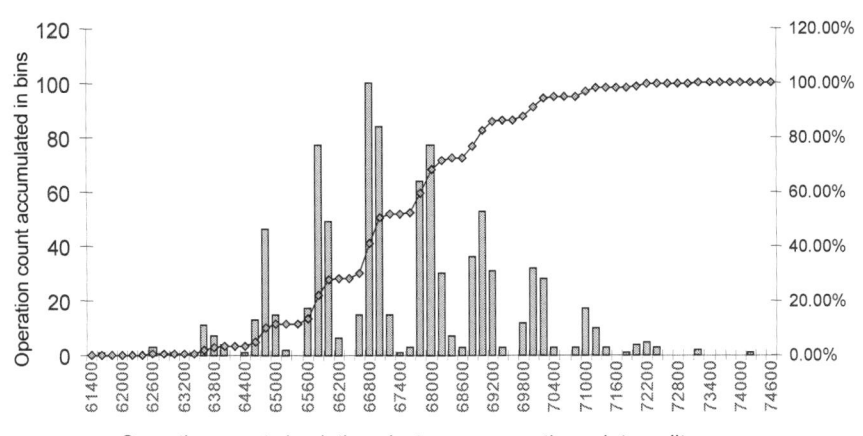

Operation count simulating electron passage through two slits

The dual slit experiment with the resulting image of the undulating envelope, which is called the interference pattern or the superposition pattern, may not seem a particularly difficult phenomenon to pursue experimentally. There are instruments that can detect the electron as it is passing by, and an experiment was set up to do just that – by placing additional detectors right in front of the slits. However, when the determination is made which slot the electron passes through, the interference pattern disappears and the electron behaves like a straight-moving particle. When the electron detector is removed the interference pattern reappears.

The phrase 'interference pattern' coined by Young contains a poor choice of a word. Forces, light, and waves in general do not interact

---

In 1947, Claus Jönsson demonstrated electron superposition with electrons going through slits. The manufacturing of very narrow slits with close tolerances was a feat in itself and some writers identify Jönsson as the first person working on microstructure scale. Akira Tonomura later demonstrated electron superposition with electrons branching around a bar. In the Japanese tradition the heavenly post or the heavenly bar has a special significance in that the first man and woman walked around the heavenly post in the opposite direction and commenced procreating – perhaps as they became inclusive in the act of superposition. The undulating formations of mountains are all creations that could be thought as arising from mutual superposition. The spear that the first man and woman used as a pointer created the largest island, and that is agreeable with the spear being the axis of symmetry. Some interaction can perturb or partially reduce the wavefunction. In the story, the woman is advised not to be too noisy, possibly to keep the wavefunction from distorting.

●●○○○○○ **Star Numbers and Operators**

by claiming exclusivity and do not interfere with each other or displace each other. Instead, light and other waves interact inclusively with resulting superposition where waves add up and stand on top of each other or subtract from each other. Superposition pattern is a better choice than interference pattern and we will use it from now on. Young also demonstrated that light's superposition pattern always retains even symmetry when he smartly introduced a shield on just one side and between the slits and the screen.

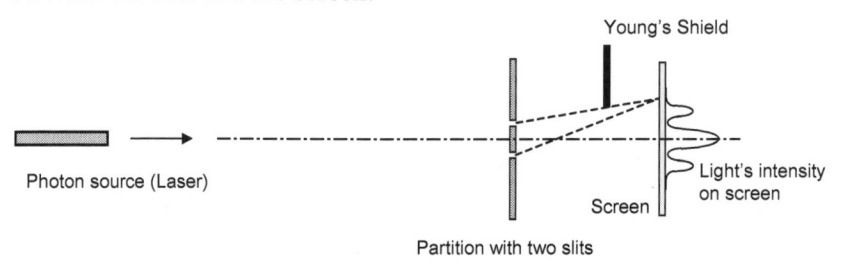

Young's Shield blocking but one side

    The resulting image shows that the superposition pattern retains even symmetry and the image is trimmed from both sides even though the shield is placed on but one side. Young's demonstration of evenness is particularly helpful when thinking of light as a force that arises only when the conservation of momentum holds and, therefore, the even symmetry guarantees that force arises only as a pair in the form of both the action and reaction.

    An electron's vibration is computationally equivalent to its velocity because the frequency of vibration is a measure of energy of a wave, just as velocity is a measure of energy of an object that has real mass. A real object's moving energy is proportional to the object's velocity and mass. The correspondence between an electron's velocity and its vibration is none other than energy conservation at work. A real electron that is moving with a particular velocity has a certain energy that matches up exactly with a particular vibration frequency. For energy conservation to hold, an electron has either velocity or vibration. Presently, the electron gains velocity by being accelerated through a field formed by electrostatic potential. Once the electron begins to move, the electron also acquires vibration, and the corresponding wavelength could then give the electron wave characteristics.

Quantum Pythagoreans

When a moving electron starts to spread in 3D the electron enters the virtual domain and its vibration *begins to manifest* as a wave. Manifestation is a word scientists do not use often, for their emphasis on real things presumes that the manifestation is a given and things are always available for measurement. Nevertheless, the electron that is not allowed to spread is a real electron that is localized and, moving or not, a real electron cannot manifest its wavelength. In the chapter describing the complete model of the electron, the mechanism that binds the electron is discussed and when the electron is bounded the wave parameter of a moving electron exists but does not manifest. The observation of wave behavior can happen only for the unbounded electron, also called the free electron.

The formation of the superposition pattern is not commutative and the relativistic presumption does not hold. The energy imparted on to the electron, or any other particle, stays with the particle and the frame of reference is not arbitrary. The electron's vibration is commensurate with its energy and it is not possible to mathematically maintain the reality by changing the frame of reference. For example, if the frame of reference is moved onto the previously speeded up electron, the electron's wavelength becomes, mathematically speaking, zero, and the computed superposition pattern would not match the actual pattern observed in a laboratory. The relativistic postulate says that the frame of reference can be moved to any spot without altering the fidelity of the mathematical simulation of reality. But that is not the case with the dual slit experiment. However, since the energy imparted onto a particle stays with the particle, a particle having no wavelength is at absolute rest. The absolute rest can indeed be determined: When a free particle produces no superposition pattern at the dual slit, the particle is at absolute rest.

When the electron transitions from the real to the virtual domain it becomes the virtual electron. The virtual electron has vibration that is commensurate with its velocity. The measurement of electron's velocity or its vibration – that is, its wavelength – is a measurement of the electron's moving energy. The measurement of either the velocity or the wavelength works. The measurement of either parameter results in the measurement of an identical amount of energy because energy is conserved.

●●○○○○○ **Star Numbers and Operators**

Real electron with vibration and velocity [de Broglie]

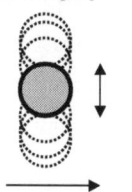

Free electron spreads and its position becomes uncertain [Heisenberg]. Electron becomes virtual electron. Vibration begins to manifest as electron wavelength

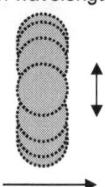

Moving electron becomes virtual electron while maintaining energy equivalence

Moreover, the virtual electron's vibration cannot be seen as a classical back and forth vibrating movement of electron mass in space. There is no mass parameter associated with the measurement of a wavelength; energy is computed based on the wavelength measurement alone. Suddenly, parameters such as mass, velocity, and position do not matter. Werner Heisenberg formulated some of the electron's parameters and went a step further. Not only is the velocity and position not needed to keep track of energy, but also the velocity and position cannot be known with certainty. The parameters of position and velocity are released from formal rules and these parameters now become subject to other variables. Nevertheless, the conservation of energy holds at all times because the virtual electron has a definite and *certain* vibration frequency (wavelength) while the parameters of reality such as position and velocity and mass are not necessary in conserving the value of the virtual electron's energy.

There indeed exists equivalence between the electron's velocity and vibration but the question calling for answer is: if the vibration value is certain, how could the velocity *become* uncertain? The energy of the electron is conserved as vibration when the electron is transforming and becoming the virtual electron. During the transformation to the virtual electron the electron is spreading and its velocity is becoming uncertain but the electron's energy continues to be conserved through its vibration. There can be and there are no loose ends when it comes to the conservation of energy. The conservation of energy is the only link

Quantum Pythagoreans

assuring the integrity of the transformation process and energy must be accountable and therefore certain *at all times*.[30] With the energy being conserved through vibrations, other parameters that are not uniquely linked to energy become free to transform. Asking for a definite electron velocity is asking for the real electron and after it transitions back from being virtual to being real. The issue is not the measurement of velocity but of energy. Electron velocity measurement in and of itself does not measure energy unless and until the electron has real mass moving at a definite velocity. Electron's energy is always conserved because the virtual electron's energy can and does manifest as vibration. Alternately, a real electron's energy manifests as velocity that, leveraged by mass, is momentum. An electron spreads and its velocity becomes uncertain while the conservation of energy holds. An electron can become bounded and its velocity can become certain while the conservation of energy holds. The conservation of an electron's energy holds whether the electron is spread or not. Heisenberg was taken aback after his discovery of position uncertainty but after a while he focused on frequencies in his computations and looked for variables that could be measured after the transformation. Heisenberg constructed and applied his matrix computations to the parameter of frequency that is energy-wise invariant during the transformation process. Before the transformation the velocity can be measured while frequency is computed. After the transformation the frequency can be measured. Because the transformation process between momentum and frequency is reversible, the real mass of the electron can transform between the two domains while the mass of the electron is not destroyed.

Classical scientific method objectively describes a phenomenon by determining its repeatable and quantifiable characteristics. The objectivity of the present, or classical, scientific method arises from using measurements that verify the hypothesis and confirm theoretically obtained equations. There is great comfort in the ability to describe, by

---

[30] The enforcement of the conservation of energy is through computability. Electron transformation can then be visualized as a tractable interaction between electron environmental parameters and electron's own parameters. Tractability enforces energy conservation.

95

equation, any phenomenon. In the case of a single electron simultaneously passing through two or more slits the confirmation by measurement does not work. Not only does the measurement spoil the experiment, but also there is no hope of understanding the superposition phenomenon with the measurement at the entrance to the slits because the measurement inherently offers but a localized answer. When an attempt is made to verify the hypothesis by the yes-or-no measurement, another phenomenon replaces the first phenomenon. The measurement shifts the experiment into a qualitatively different context which has very little to do with the initial observation. The premise of 'under controlled conditions' forces the experiment to remain in one domain and excludes the experience of the other domain. There is, nonetheless, nothing mystical about this: The measurement has priority over the spreading of the virtual electron.

While a single free electron indeed passes through both slits simultaneously, we can only arrive at this conclusion indirectly. Welcome to the cave of the shadows and the phenomena of transformation. Welcome to circumstantial methodology and correct conclusions obtained by preponderance of circumstantial evidence. In the end, a virtual electron's position continues to be computable with wave mechanics while the virtual electron's spread is the electron's position parameter that is now distributed in space.

The electron must spread since it is exiting both slits simultaneously and superposes with itself after it leaves the slits. The advance of quantum mechanics to the macro scale now becomes possible. To see this electron again means that the electron's position becomes known and therefore there must be a way of reducing the electron back to a single spot. Shining light on the electron is but one way of doing it. When the entire dual-slit experimental setup is exposed to light the electron superposition pattern disappears. Light accomplishes the same result the electron detector did when the detector is placed at the slits.

Once the illuminating light's wavelength is short enough, high positional accuracy of the electron can be obtained and the electron does not and cannot spread. It is now an opportune time to introduce the concept of *bounding*, for the virtual electron as a whole can be bound in a larger or smaller spatial volume. Shorter photonic wavelengths bound the virtual electron in a smaller spatial volume that is proportional to the photon's wavelength. Once the electron is tightly bound the position of the electron is known to high accuracy because the extent of electron spreading is near zero. The superposition pattern disappears when light bounds the electron to a size that is smaller than the distance between the two slits. The important observation here is that light's wavelength has priority over the spread of the virtual electron and light does not bounce or scatter from a free moving electron. Light does not knock free electrons out of their path if there is no exchange of energy between photon and electron, which is indeed the case. Virtual electron conforms to light, including visible light, and the electron is not in light's way. Light's frequency (light's color) does not change in the encounter with the virtual electron although its speed will slow down somewhat during the interaction and that results in the phenomena of refraction.[31] Virtual electrons are "invisible" much the same way glass is invisible. In the interactions between light and a free moving virtual electron, light's wavelength is the independent variable whereas the electron's spatial spread in the virtual domain is the dependent variable. This has yet another and very far-reaching aspect: Matter may disappear by exiting the real domain but matter cannot escape detection by light, and matter cannot break out from the dominance of light.

In the dual slit experiment the electron detector, when placed in front of the slits, binds the electron fully by the act of measurement and that is why the superposition pattern disappears. Another way of saying it is that a tightly bound electron behaves like a classical well-defined particle although it remains classical only if the free electron is actively bound. Electron bounding by the act of measurement also makes logical sense: If the electron is detected here then the electron cannot be there because the electron cannot be physically split. When a free moving electron is not being detected or exposed to light, however, it spreads out and is at different places simultaneously because it exists in a spatial volume that is greater than the real electron's physical size. When the virtual electron's spread is greater than the distance between the two slits

---

[31] Shorter wavelength of light has greater refraction in its optical interactions with the electron.

the superposition pattern starts to form. Once the electron encounters the material screen or electron detectors on the other side of the slits, the electron becomes bound and appears as one real entity because the act of encountering the impenetrable screen facilitates the electron's position measurement.

> *To dance for our pleasures*
> *We prick the shadows to come*
> *From out of the cave*
> *Strange relations*
>
> *They'll come out for a visit*
>   *if you poke them with a boom*
> *You can pick them if you like them*
>   *under hyperbolic moon*

While electron spreading depends on the interaction with elements such as light or the physical structures it encounters, the bounding or the reducing effect is instantaneous. A free – that is unbounded – electron instantly reduces into one spot in its encounter with the screen. John von Neumann mathematically described how a wavefunction collapses instantly. Collapse is von Neumann's term but the term *reduction* is beginning to replace it, particularly since the reduction can be partial or full. Wavefunction is another word for a particle's virtual existence and can be visualized as positional distribution of the virtual particle – that is, the particle is spread out in the virtual domain and the wavefunction mathematically describes the particle's spatial spread. There has been much discussion on the interpretation of the wavefunction and several views are still active. Mathematically, the shape of the wavefunction exists and varies in space but is the particle itself spread in space? The particle is indeed spread in space while the particle's real parameters transform into virtual parameters.[32]

Von Neumann's mathematical discovery of the instantaneous reduction of the wavefunction is likely the most radical departure from classical physics. The instantaneous nature of the reduction together with

---

[32] Virtual numbers are based on a square root of minus one. At times called imaginary, virtual numbers will be described further in the upcoming chapter.

Quantum Pythagoreans

Paul Dirac's mathematical treatment of antimatter is the defining hallmark of quantum mechanics. In view of the fact that discontinuous movement on a macro scale relies on the mechanism of instantaneous reduction, von Neumann's discovery is in the most remarkable category.

Schrödinger's wavefunction equation describes the virtual electron spreading in space. This equation is similar to another equation that describes the diffusing penetration of one thing into another. Schrödinger's equation has virtual diffusion coefficients while the classical diffusion equation has real diffusion coefficients. Electron's spreading in space is described by a time-evolving wavefunction that can be also likened to a diffusion of the virtual electron into space. Electron diffusion – or spreading or evolving – is independent of the speed of the electron and so the electron is "dissolving" whether or not it is moving – while carrying vibration that is commensurate with its energy.

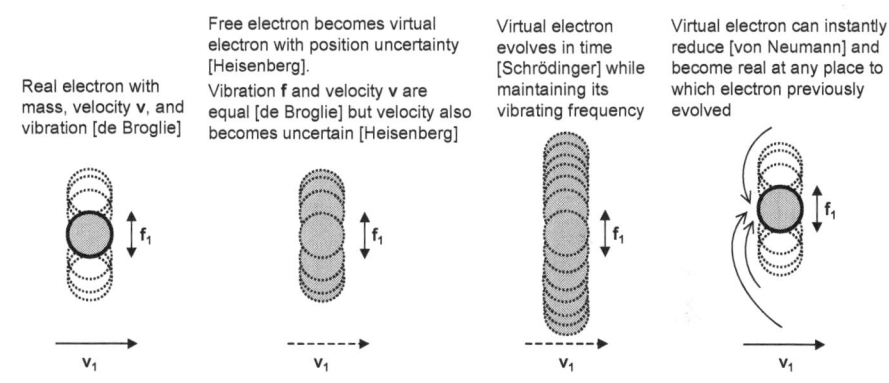

Two energy equivalent states of a moving electron with evolution and reduction

Outside of the atom, a free and moving electron can and does spread while being able to instantly reduce partially or fully. The act of electron measurement presents the electron with a yes-or-no construct and the virtual electron's reduction results in a real electron's reappearance in one spot. However, when a free and moving electron encounters two slits, the electron does not reduce completely but propagates through both slits as it maintains its computable and virtual state. Two slits or other geometries, then, present the virtual electron with various computational constructs. An electron's computational interaction with structures it encounters creates an unlimited number of various wavefunctions, for dual slit interaction produces just one family of wavefunctions. When the virtual electron encounters different geometries the electron computes a wavefunction for that geometry and such computational interactions present a great opportunity for a new class of computing.

Much has been said about measurement and what measurement means in the quantum mechanical environment. It is indeed possible to go beyond giving the measuring instrument the imperative of enforcing the: "Are you there? Yes or no!" question. Electron spreading and its reach, for example, is the dependent variable of the electron while the unbroken physical composition of a screen is the independent variable of matter's spatial position. The transformation into the real domain then allows the measurement of a magnitude of a real parameter such as velocity or mass.

In summary so far, a stationary or moving electron spreads in a way that can be visualized as diffusion, which is in accord with Schrödinger's wavefunction equation. Electron spreading enables the electron to pass through many slits simultaneously – that is, without reducing in full. When the electron is moving it acquires vibration frequency and this allows the electron to behave as superposing waves after passing through a plurality of slits. The vibration frequency is absolute and only the energy actually imparted on the electron will manifest a particular wavelength – a change in the speed of the frame of reference has no effect on the electron's vibrating frequency. A virtual electron spreads in space and its energy is conserved as vibration frequency. An electron spreads in space and its parameter of velocity is uncertain but its vibration remains certain. Once the electron velocity is uncertain, electron momentum is uncertain as well. De Broglie described electron's vibration by mathematically relating momentum to frequency, for the vibration frequency is energy just as momentum is (a moving) energy. The real punch of the real electron's momentum becomes the virtual electron's wavelength that can be verified by measurement. Higher electron speed gives the virtual electron a higher vibration frequency, which manifests in a tighter superposition pattern that can be computed and verified.

A free moving electron can be visualized as a single volume entity that instantly adapts to its environment. A moving virtual electron is a meld of spatial volume and vibration energy, and together these determine the computational result of interactions. A moving virtual electron has independently varying spatial volume and frequency components. In addition, a virtual electron's spatial volume evolves (increases) in time and in accord with the Schrödinger equation. The result of the virtual electron's interactions with the environment depends on both the spatial geometry the electron encounters and its vibrating frequency. A moving virtual electron is a moving packet of space-energy continuum. It cannot be parted into separate and independent spatial volumes but can pass simultaneously through many slits if the virtual electron can remain within one contiguous spatial volume. The shape of the moving virtual electron changes instantly and can be pierced or pinched instantly. Overall, the reduction in spatial volume of the virtual electron is instantaneous but the increase in virtual electron spatial volume is subject to the Schrödinger equation's time-dependent evolution parameter.

If the virtual electron fully reduces it becomes a real electron and at that point the real electron's energy instantly manifests as mass and velocity. The instant manifestation of the real electron in one spot enables the measurement of either the wavelength or the velocity and both measurements are successful because either measurement is a measurement of the electron's energy. To verify a real electron's velocity, the measurement of but one electron is needed. To verify the wavelength of the electron, however, requires many measurements. Reduction of the virtual electron happens anywhere along the virtual electron's reach (spread) that exists just prior to the measurement. The electron does not reduce repeatably into one and the same spot and the verification of the electron's wavelength requires full reduction of many electrons. As soon as the electron instantly reduces to its classical form, and if the electron remains a free electron, it begins to spread again. If such electron is moving, it then acquires vibration as well. A free electron becomes a real electron for an instant at the instance of measurement, and spends the rest of its time as a virtual electron.

The wave-particle duality continues to be a much-discussed topic, as it deals with one of the high points of early quantum mechanics. At least three physicists are reported as saying something akin to 'nobody understands quantum mechanics.' They certainly spoke for themselves, for none of them offered an explanation. The salient point of the wave-particle duality is that the physical particle and the *manifestation* of the particle's wavelength are mutually exclusive and that means that the electron is never a real and a virtual electron at the same instance. The virtual electron interacts with the environment as a wave. But to confirm wave behavior, many repeated measurements need to be performed, for each measurement yields but one position, or a point, of the wave. Electron reduction at measurement is instantaneous and after many measurements the wave shape emerges. Electrons are detected forming a pattern and, in an environment with symmetrical geometries, the pattern can also be calculated and forecast with wave mechanics. The electron wavelength is certain and the dual slit superposition pattern is the most quoted example.

Electrons deflect at a particular angle when directed at crystal lattice – the same way waves of light of identical wavelength deflect. A crystal lattice consists of rows and columns of atoms with much space in between; this arrangement favors particular reflecting paths. The crystal, then, does not necessarily force the electron wave to reduce. Reflections from a crystal superpose and add up to some maximum at which the electrons are detected and, consequently, the angle of reflection may or may not be the same as the angle of incidence. Because the virtual electron instantly reduces upon measurement the experimenter then also measures electron velocity and so the electron instantly becomes a real particle with mass and velocity. A free electron, though, always becomes a virtual electron while becoming a real electron only for the duration of the measurement or active confinement.

Yet the most unusual aspect of the reversible wave-particle transformation is that a moving particle always has vibrations. The particle must be free if it is to spread and be able to manifest its vibration but, bounded or unbounded, a moving particle always carries vibration that is commensurate with de Broglie relation. Vibrations can be linear and angular, and the full extension of de Broglie vibrations is forthcoming in the chapters on the virtual domain.

Branching and Spreading

The interferometer instrument presents a beam of light with a tilted half-silvered mirror that "splits the beam." A portion of light's energy is reflected while the remainder passes through the mirror. The interferometer enables the measurement of a difference between two spatial distances because each portion of light first travels the desired distances and both portions of light are then brought together. Once brought together the time difference in crest arrival determines the difference between the two distances each of the crests traveled. It is now possible to see that each and every photon branches instantly into two arms at the half-silvered mirror. The reflected arm and the passing-through arm now each have one crest and both propagate in different directions. Every single photon branches into two arms, each arm of the photon expanding at absolute speed which is lightspeed. If, say, some photons pass through the mirror as individual photons and if the mirror later reflects other photons as individual photons then the crest shift could not be linked to travel distances when branched beams are reunited.

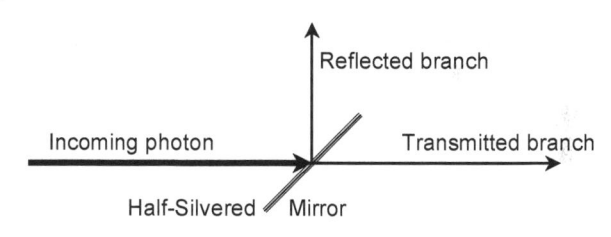

Light beam "splitter"

Each and every photon indeed branches simultaneously into two arms – one half-size wave crest in each branch – but because the photon cannot be split into a plurality of independent sub-photonic entities, both photonic arms are interconnected through the branching point, thus forming a single wave where both crests of the wave propagate at lightspeed in different directions. The interconnectedness of a branched photon forms the basis of light's nonlocality that is truly a magical property. The upcoming chapters address nonlocal light behavior. It may be difficult to imagine a link existing and stretching on without limit, but both crests of a branched photon are indeed interconnected through a link that has no distance bounds. In essence, a photon's energy is fixed and the conservation of energy holds computationally without the parameter of distance. The absence of the parameter of distance can be seen as limitless distance or as distance being irrelevant. The conservation of energy is based on computability, and the energy of a photon can be computed without the parameter of distance. Computability, then, is at the heart of energy conservation. Additional discussion on computability will be addressed in the chapters on gravitation.

The interferometer's half-silvered mirror presents the photon with a branching construct similar but not identical to the geometry of the dual slit. In the interferometer the crests of the branched photon each propagate in a straight line with one degree of freedom (1D).

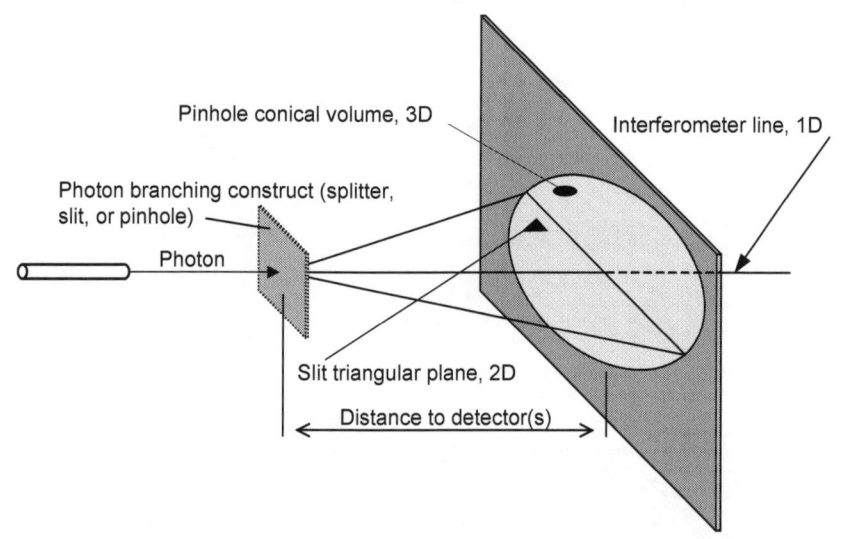

Depending on construct, photon branches with different degrees of freedom

Coming out of the dual slits, each of the branched ends of the photon propagates in concentric circles of the plane perpendicular to the slits and thus each photonic end propagates with two degrees of freedom (2D) of the triangular plane: When slits are vertical, the 2D of propagation freedom is in the horizontal plane. If photons were parted with dual pinhole geometry, then coming out of the pinholes each of the branched ends of each photon propagates in concentric spheres and thus each photonic end propagates in the conical volume with three degrees (3D) of freedom. In all cases the photon is branched with both ends, each end having one, two, or three degrees of freedom while both ends remain interconnected. After dual slit branching both photonic ends superpose in 2D. In the interferometer both photonic ends (crests) superimpose in 1D if both ends are brought together after traveling nearly identical distances.[33]

A photon encountering a pinhole and a vertical slit right next to each other would branch such that the photon end passing through the pinhole will spread with three degrees of freedom while the photon end passing through the slit will have two (horizontal) degrees of freedom. The resulting self-superposition will produce interaction between a 2D plane and a 3D cone – a hyperbolic superposition. Because the electron partitions and spreads with the same mechanism as the photon, similar superposition arrangements can be made with the electron as with the photon. The electron's wavelength depends on its speed so its wavelength can be tuned to any frequency.

The electron as well as the proton are atomic components that are considered irreducible because they cannot be broken up by force or because they do not break out to other particles. The irreducibility, however, should not be taken to mean that the electron could not spread, stretch, or become doughnut-like or have a shape of an hourglass. Actually, an argument can be made that the electron is shatterproof because it can yield and spread – and regroup when influences change. An electron remains a whole entity when it adapts and changes its wavefunction shape as the electron interacts with the structure inside the atom, for example. The wavefunction not only represents the electron's virtual presence in space – it is the electron's computational presence and computational reach in space. The instantaneous reduction of electron wavefunction materializes the electron in one spot and the particular positional appearance is in accord with the electron's wavefunction the time of reduction. The wavefunction magnitude is then also the probability of the electron's positional appearance.

---

[33] Branched photons have crests of finite duration. While both crests are interconnected the magnitude of the wavefunction between crests is very nearly zero. Once the photon is branched and if one crested part gets ahead of the other by traveling a shorter distance, the photonic crests cannot superpose with each other. Both crested parts need to travel nearly identical distance to superpose with each other.

●●○○○○○ Star Numbers and Operators

Depending on photonic energy inflow or outflow in or out of the atom, the electron can instantly materialize on a level that keeps the atom in computational balance. The computational balance is achieved at a particular distance from the core where the real electron's energy equals the virtual electron's energy. The real electron's energy is a multiple of its mass and velocity while the virtual electron's energy is its wavelength that, in turn, is related to the orbital frequency. The electron's energy is added to the energy of the impinging photon and if the net energy adds up to an available orbital, then the photon is absorbed and the electron moves to a new orbital.[34] Electron computations and transitions within the atom can be also likened to the enforcement of energy conservation when the atom is subject to external stimuli. Real electron momentum – or electron's moving energy – has transitional equivalence in the vibration frequency of the virtual electron. A real electron is a near zero-dimensional point with real mass and its speed determines its moving energy. A moving virtual electron is a vibration of a moving and contiguous volume of space and its vibration frequency determines its energy. If you reached the conclusion that the real electron has real energy while the virtual electron has virtual energy, you are doing well. Schrödinger's equation includes virtual numbers but de Broglie's math is sloppy in this regard.

A free moving electron can increase its energy by increasing its vibration – that is, virtual energy in the form of vibration can be used to add to the electron's energy. If the electron becomes a real electron later on, the conservation of energy then guarantees the real electron to

---

[34] Specifically, a real electron's energy is a product of its real mass and angular velocity. A photon's absorption is in the framework of momentum conservation and one half of the photon's energy is added to the electron's energy in the search for a new orbital. An electron becomes a real electron for but an instant and acquires real parameters in an eigenstate (coming up). Orbital availability is linked to incomposite (prime) numbers studied extensively by Georg Riemann, who also adopted "his own" non-Euclidian geometry to accommodate atomic orbitals. Riemann geometry is custom tailored to atomic geometry. For example, all intersecting orbital paths form triangles that always have the sum of internal angles exceeding 180 degrees – in accommodation of the spherical geometry. Also, all paths – that is, orbitals, are finite in length and in deference to circular paths that always repeat and are not, therefore, unbounded.

Quantum Pythagoreans

instantly materialize with higher velocity. While the increase in electron velocity has not been confirmed with controlled experiments to date, the increase in velocity is predicated on the conservation of energy and is shown below.

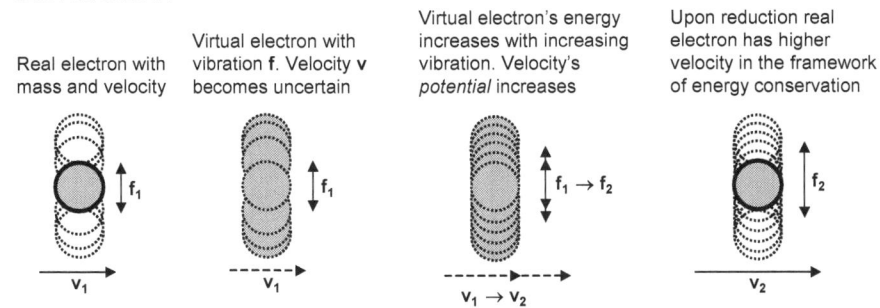

Real electron with mass and velocity	Virtual electron with vibration f. Velocity v becomes uncertain	Virtual electron's energy increases with increasing vibration. Velocity's *potential* increases	Upon reduction real electron has higher velocity in the framework of energy conservation
f_1 v_1	f_1 v_1	$f_1 \rightarrow f_2$ $v_1 \rightarrow v_2$	f_2 v_2

Virtual electron's increase in vibration frequency converts to higher velocity at reduction

    The prerequisite is an increase of a particular electron's vibration such that upon its reduction the real electron just jumps to higher velocity. This mechanism accounts for presently unexplained electrostatic motors and forms the foundation of the phenomena of ball lightning. The facility that increases or decreases the virtual electron's frequency could explain sudden if temporary failure of electric and electronic appliances, computing machines and neuron processing. The ability to change a virtual electron's frequency directly and at will is certainly a major technological gateway of the 21st Century. It is likely this mechanism will be applied in the explanation of psychokinesis and could leverage it to higher energies. Because the frequencies can be decreased as well as increased, yet another potential application can be had in the ability to stop moving bodies at a distance. De Broglie vibration applies to all moving bodies. The conservation of energy holds and to stop a moving body at a distance an appropriate amount of energy needs to be expended, but the form of energy being applied is now different.

    The indestructibility of the electron hints at a form of intelligence that sustains the electron by its ability to repeatably transition between the real and virtual domains while interacting with the environment in two separate computable modes. A particular electron is either real or virtual but never real and virtual, for the computational needs are different in each domain.

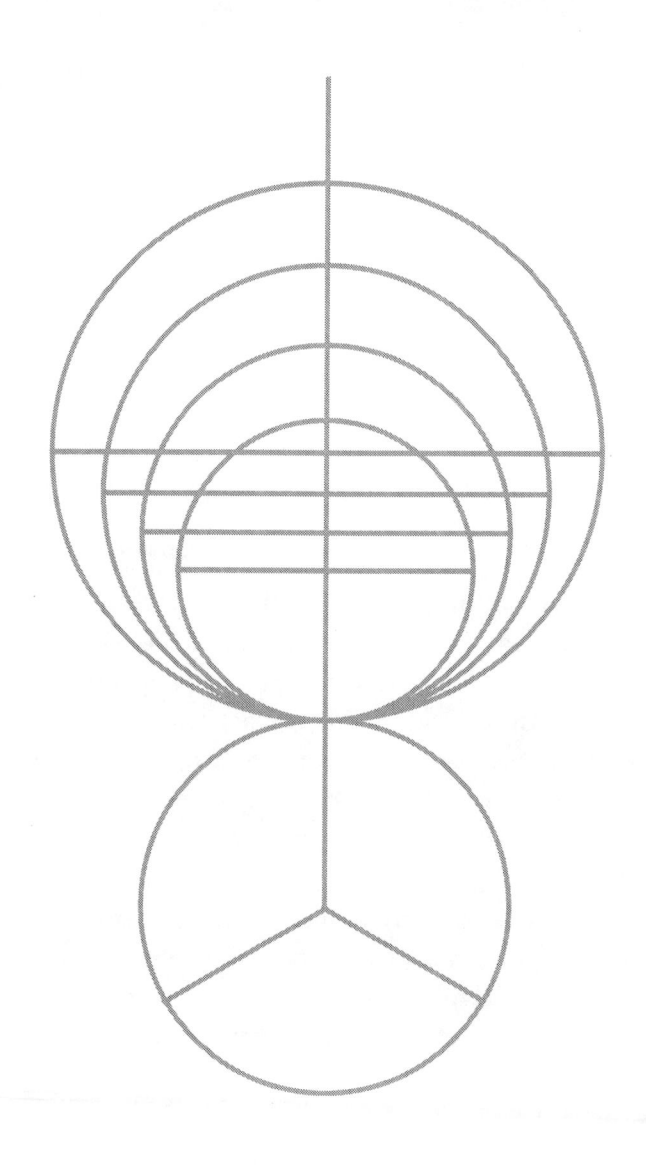

VIRTUAL ELECTRON	110
Electron's Virtual Vibrations	*116*
Virtual And Irrational Numbers	*122*
Complete Model of the Free Electron	*138*
FIRST CONJUNCTION. THE ATOM	144
THE VIRTUAL NATURE OF LIGHT	151
Unbounded	*156*
Momentum Arises	*157*
Heat And Pressure Mechanics	*159*
Light Mill	*164*
Separated But Undivided	*166*
More To Lightspeed	*171*
Photonic Energy Coming and Going	*173*
Self-Superposition	*178*
The Planck Constant	*181*
Superposition of States	*187*
Light, Just Warming Up	*191*

The Virtual Domain

You will not be able to see the shore until you shore the sea

    The cave of the shadows metaphor is pretty much used up. The shadows are now going to dissipate some of their connotations. There are other concepts that, while formidable, are different rather than strange. A free moving electron spreads and becomes the virtual electron because its classical parameters of position and velocity acquire virtual and nonlocal characteristics. Position and velocity are now free to acquire many possible values – a multitude of values – because virtual parameters are nonlocal. The virtual electron now also has new and different properties that mesh with the virtual domain. Electron spreading and its instantaneous reduction is something very special and to some it is close to magical.

    Light's energy is fixed for the life of the photon. What appears to be a limiting characteristic actually enables the photon to stretch and spread without bound. The energy of a photon is always computable and the conservation of energy holds without regard to distance.

Virtual Electron

Thomson, de Broglie, Heisenberg, Schrödinger, von Neumann and Dirac (coming up) all made significant contributions to the understanding of the electron. While most physicists can deal with classifying the electron as real or virtual, the next obvious yet difficult step is that in effect and in fact the real electron becomes the virtual electron by dissolving into space. Disappearance of an electron's mass is partially clarified by Heisenberg by way of uncertainty. The position of a free moving electron, Heisenberg discovered, *cannot be* known with certainty and this is yet another way of saying that the moving electron is spread in space. By comparison, any real object that is vibrating fast or slow is at any given time in a certain, definite and measurable position. Heisenberg thus describes the virtual electron, which has a virtual rather than real presence. What is also becoming apparent is that the vibration of the real object is not the same as the vibration of the virtual entity although both kinds of vibrations are expressed with frequency of cycles per second. De Broglie spoke of probabilistic 'matter waves' but again he is not explicit about the electron shedding its classical parameter of mass.

Heisenberg arrived at the uncertainty relationship by working the de Broglie equation. By becoming a wave, the electron is no longer localized and electron spreading becomes analogous with statistical distribution. The answer to "How many people in the world are exactly six feet tall?" is zero. There are many people shorter and taller than six feet but there is no one who is exactly six feet tall. When the electron is spread in space its distribution is similar to the population distribution, and an electron's probability to be at an exact spot is zero. Just as with statistics, the answer involves the integration of an area and, therefore, the position in which the spread-out electron could be found must include the range of spatial distance.[35]

---

[35] Because of Tetractys, Pythagoreans are for the most part comfortable with the notion that what works in 2D (works for an area) does not necessarily work for a zero-dimensional point. The area under a section of a curve may be computable through integration but asking for an area under a point results in zero. Every day there are people who go to bed shorter than six feet and wake up longer than six feet – so they "must" have been exactly six feet at some (one) point. Yet the point

Quantum Pythagoreans

Yet there is more to Heisenberg uncertainty than spatial position alone. The Heisenberg uncertainty principle trades off electron position versus velocity and so the exact knowledge of velocity excludes the exact knowledge of position and vice versa. The measurement of the electron reduces the electron instantly and so the velocity (or momentum) of the electron is known instantly and exactly, but because the electron is spread out as a wave *just prior* to the measurement, the actual position at reduction is not known ahead of the actual measurement. Electron position at appearance (at measurement) then has a particular 1D, 2D or 3D spread and hence the electron has position "uncertainty." When the electron is in its virtual and spread-out state, it has position uncertainty because it is nonlocal. At measurement the electron reappearance has position uncertainty because the electron appears anywhere along the reach of the virtual electron and its position at appearance cannot be forecast. While it may seem that the inability to know ahead of time where the electron will appear is some form of a deficiency, it is just this characteristic that enables the reconstruction of the wavefunction of the electron. Many reduced electrons begin to form in a pattern that reflects the probability *and* the reach of the spread of the virtual electron.

If the experimenter needs to measure the electron's exact velocity, the electron's position at appearance is likely of no consequence. But if the experimenter calls for an electron's reduction to happen in a particular small spot then he or she has to wait until some electron actually happens to reduce in that very region. The waiting time may be short or long, which then gives the mathematical result that the electron has a range of velocities because of the range of arrival times. But enlarging the region in which the electron is to appear also narrows the spread in waiting time and in calculated velocities. If the spot in which the electron is to appear covers the entire wavefunction, then the electron appears instantly with exact and repeatable velocity somewhere along the reach of its wavefunction.

The science of physics tends to bias toward dealing with physical and tangible reality. Some physicists would rather deal with alternate realities instead of a single intangible and invisible pocket of vibrating

---

in time in which a person is exactly six feet is indeterminate because it has an infinitely large mantissa.

111

energy that is the virtual electron. Other physicists postulate multiple universes so the electron can remain real in at least one of them. When the electron disappears, disbelievers can be placated by the theory that the electron did not actually disappear because it is in 'another real universe' with all of its real attributes. Some physicists plunge into intractable analyses of all possible position histories in an attempt to keep track of the disappeared and diverging or superposing electron. There is even one back seat driver theory that proposes to wait for all uncertainties to settle down before – but only in retrospect – one can describe what happened. Some reality defenders advocate a do-not-go-there attitude because the inability to deterministically describe what is happening in every conceivable instance of time means that somewhere somehow there is something missing. But even if being real at all times has appeal, there are disadvantages. Real systems can be or can become chaotic. Chaotic systems, in turn, are intractable. Yet, the chaotic system is completely real and each and every body's position is known with arbitrary accuracy through measurement. Moreover, the traveling salesman problem does not happen in a chaotic context but it is an example of a real yet intractable problem that cannot be solved in a practical timeframe.[36] Chaotic systems cannot be described with real methods in real-time and, therefore, real models do not and cannot begin to deal with chaotic systems. Reality cannot banish intractable chaotic behavior and any system that is or

---

[36] A salesman needs to travel to many cities and wants to minimize the total route. The only way to pick the shortest route within the attributes of the real domain is to exhaustively calculate all possible routes first. The traveling salesman problem is classified as nonpolynomial, which puts it in a category of intractable problems. The number of all possible routes **r** is computed by taking **n**-factorial, also denoted as **n!**, where **n** is the number of cities. The number of all unique routes **r = ½ n!**, where **n! = n·(n-1)·(n-2)... ·2·1**. For ten cities, for example, the number of all possible routes is approaching two million. **69!** is a number on the order of 10^{99}, which is a number much, much greater than Archimedes' *Sand Reckoner* estimate of the number of the grains of sand that, at 10^{63}, would fill up the entire universe.

Newton is the author of factorials-based infinite series that converges toward the value of π. Because the factorial is in the denominator and grows with each term, the sum yields additional significant digits of π very quickly.

becomes chaotic cannot be fixed with real methods – short of destroying it in some way. The inclusive concurrency of the virtual domain, however, can address intractability and, therefore, chaos. If the traveling salesman were to enter the virtual domain he would not have precisely defined positions but that is exactly why the salesman could then be in two or more cities at the same time. It is the wave with its nonlocal property that spans spatial distance as one entity. Symbolically, the wave is oftentimes described as wings or feathers, although feathers have an additional memory component due to their geometry.

Alternate or multiple realities do not help in the visualization and applications of the virtual electron. The virtual electron is in the virtual domain rather than in some "other reality." Virtual electrons exist but are no longer subject to the reality of mass, for example, while their energy is that of vibration rather than velocity. Reality is about exclusivity. In the real world or in the real domain there is but one house on one foundation and every fender-bender is a reminder of what reality is about. Reality is about objectivity because a formula working in one lab works in all labs. There is no such thing as subjective reality. Over a course of a particular event people may have had different perceptions of what happened but they are remembering different subsets of facts of one reality rather than different (and subjective) realities. However, the ability of the virtual electron to superpose gives a strong indication that the virtual domain is about inclusiveness rather than exclusiveness.

The virtual domain is a realm where virtual entities move, evolve, superpose and interact. Technically, virtual entities do not need a medium in which to move, for they move or stretch whenever they interact and virtual entities always interact. Virtual entities exist in their own right and again, technically, they do not need a substance that would mediate their interactions and they do not need some such medium for their existence or sustenance. Virtual entities do not represent knowledge or energy because they *are* knowledge and, therefore, they are organized energy. There is no such thing as neutral knowledge that would not relate to anything else. The virtual domain consists of virtual, that is – intangible – entities that interact among themselves in a way that is different and unique from interactions among real entities. The electron is one entity that can readily become

113

The Virtual Domain

either real or virtual. Electrons can be real or virtual entities with real or virtual attributes.

The absence of mass indicates that in the virtual domain entities do not and cannot butt heads or twist arms in the literal sense. The virtual domain, then, is about relationships where the result of interactions depends on the values variables are carrying. Strong values of one variable may result in one outcome while the same values with a third variable result in another outcome because the strength of the virtual variable is always relative. The most distinguishing trademark of virtual variables is that they interact among themselves with the values of the entire variable. Temperature and pressure, on the other hand, are real variables because values of these variables issue from real entities – the localized back and forth vibration of real particles. The values of temperature and pressure start at (near) zero and can be applied as a particular measure of the variable that is the magnitude of the variable. A chemical reaction will always commence at one particular magnitude of temperature and pressure, for example. For virtual variables, however, the independence and thus the strength of a variable is always relative and the outcome of the interaction depends on the entire range of values the variable contains – that is, virtual variables interact with their values as a whole and so the entire wavefunction is always engaged. Real variables interact with individual values variables have in a particular instance but virtual variables interact with all of their acquired values in one operation.

What works in one relationship may not work in another. The order of interaction also becomes important. Asking question A before question B results in an outcome that may be quite different when the order of the questions is reversed. Yet both the real and virtual entities share the same space with an absolute distance and time grid overlay even though each treat the overlay differently. Real and virtual entities also interact with each other in an outcome that may under some conditions transform real entities into virtual entities and vice versa. Real and virtual domains are not isolated from each other by some third party. Real and virtual domains are separated from each other because they cannot become unified in some common domain. If a real entity transitions to the virtual domain it then becomes a virtual entity possessing virtual parameters while performing virtual operations in absolute space. Because of the transformation, an

Quantum Pythagoreans

entity's parameters transform as well when moving between the two domains. Exclusivity is not freely available in the virtual domain just as inclusiveness is not a given in the real domain.

Although virtual entities exist in absolute space the parameter of spatial distance is a dependent variable and distance is thus subordinated to other parameters. Another way of describing the virtual domain is that it is an informal system while the real domain is a formal system. The virtual entity not only lacks the reality of mass but also is not subject to real parameters such as temperature and pressure. A virtual entity can descend to the surface of Venus without encountering the tremendous pressure of the Venutian atmosphere. Being inclusive, a virtual entity can pass through mass if it does not computationally interact with it. Certain photonic shapes will interact with matter in the manner of absorption, reflection or refraction. Some photons, though, will pass right on through matter if their particular shape does not reduce, while slowing down or refracting a bit during interaction.[37]

Shapes and forms dominate in the virtual domain because shapes and forms steer relationships and facilitate computability among virtual variables. Figuratively speaking, shapes and forms is the medium of the virtual domain. Shapes and forms correspond to the words *physique* or *posture* as well as *geometry*. Additionally, all sciences recognize the intangible component of knowledge. The exploration of physics, then, is about the real as well as the virtual aspect of the universe. Each domain requires a unique perspective. There is a tendency to specialize and deal with only one domain. Be that as it may, the ability to work in both domains – and working with their respective methods – provides additional benefits; the increase in organization is one of them. The real and virtual methods are mutually exclusive but both methods can and should be

---

[37] In the Hindu mythology the real domain is at times referred to as Maya or "illusion." The virtual entities can indeed pass right on through matter as if matter were not there, but only if the virtual and the real entities do not computationally interact. Certain geometries, though, will present a computational environment to the virtual entities, at which point the virtual entities will interact and possibly transform from virtual and into the real domain. Some virtual entities may have no choice but to reduce into a real domain upon encountering geometries composed of real matter.

applied at different times to produce better and better understanding of our environment. Improved understanding of the environment's properties and behavior then results in improved organization of the thinking process. Working with real and virtual methods increases assessment speed and predictive power, which can be summarized as an increase in organization and decrease in entropy.

Electron's Virtual Vibrations

A virtual electron evolves and interacts as one entity with the physical geometries it encounters. Two slits produce a different pattern than a triple or single slit. Virtual electrons also interact with the lattice of periodically spaced atoms in crystals and, since the electron wavelength depends on its energy, electrons reflect from or refract through the crystal in a direction that depends on the incoming electron energy. The crystal, then, can act as an electron prism that separates electrons based on their incoming vibration energy. Electron's frequency increases and the electron's wavelength decreases with increasing electron speed. For example, electron's wavelength is responsible for the basic mechanism of the electron microscope. The accelerated electron acquires a particular speed that maps into a particular electron wavelength.

The fact that a free electron actually dissolves as it spreads has not been part of mainstream quantum mechanics. For eighty years quantum mechanics is relegated to the scale of the atom where electron spreading and reduction are oftentimes described as electron tunneling. Electron spreading, however, also happens outside of the atom, while inside the atom electron spreading is bounded by the interactions within the structure of the atom where electrons find computational closure. Perhaps there was no significant need to even think outside of the atom regarding electron spreading because the quantum mechanical equations work fine within the atom. All the benefits of transistors and lasers come from the atomic framework. Electron spreading does not come to the forefront in the case of the electron microscope because the electron wavelength is the locally controlled working parameter and the computed wavelength is in agreement with the microscope's experimental results. In the dual slit

Quantum Pythagoreans

experiment, however, electron spreading becomes relevant because the electron must envelop a partition from two sides.

Newton postulated the necessity of absolute spatial distance and absolute time. Absolute distance, time, as well as velocity are independent of the frame of reference – that is, the frame of reference is not needed. Lightspeed fits that nicely because there is no need to ask 'with respect to what?' and it is then possible to partition the speed of light and derive the absolute distance and absolute time from the absolute speed of light.

If you were to have a discussion with Newton on the subject of absolute reference, the conversation could happen as follows:

"If any observer can measure physical parameters with absolute results, one does not need to figure out the point of reference," you open.

"Of course, if all observers measure identical results then the reference is irrelevant," answers Newton, "which also means that my concern about not knowing where to place the reference would be moot."

"In the case of lightspeed the results are absolute and valid for all," you fill in.

"True. But what about all the real objects?" Newton interjects. "To obtain the real object's moving energy we need some kind of reference because the object's speed, and therefore energy, depends on the reference."

"Apparently so," you say. "Yet every time energy is imparted onto a real object, the de Broglie vibration changes to reflect the increased kinetic energy of the object. By measuring the de Broglie vibration of the object *instead* of velocity, all observers could measure the object's absolute moving energy. You started on that path when you called the objects that are accelerated with force as having *true* motion, in contrast to motion as observed from a reference."

"I knew the Continental guys were hiding under a rock – but go on."

"If we could measure the moving object's de Broglie vibration," you continue, "*and* if such measurement were the same for all moving and stationary observers then we can agree on a framework

117

of absolute measurement that does not require any reference at all. After all, when the object passes through slits the superposition measurement results in a value of a wavelength that is commensurate with its *true* energy."

Newton is silent and maybe he likes it. Absolute measurement framework dispenses with all references altogether. But now you want to explore the discontinuous movement through space and so you continue:

"Sir Isaac, should a moving body disappear it will continue to have de Broglie vibrations."

"I do not see why not. Energy is conserved even if the moving body's mass becomes virtual mass. A body's vibrating energy when it was a real body continues to exist as vibrating energy when the body's mass transitions back and forth between real or virtual mass. If the body reappears anywhere in the universe you know that the energy continues to be conserved because the reappearing real body will have the real velocity that is commensurate with its de Broglie vibration."

"Finally," you say, "a real body does not need to be in absolute rest before it transits to the virtual domain because its moving energy already exists as de Broglie vibration, which then guarantees the conservation of energy *and* direction. De Broglie vibrations are virtual vibrations and can be called *vivibrations*. Vivibration is the energy that is attached to an object for the purpose of the conservation of energy. Momentum is a moving energy that is conserved through inertia, and the vectoring or the directional aspect of imparted energy is also conserved. A body can reappear far away in the blink of an eye but that should not be a problem if it has the same moving energy with the same direction. The conservation of momentum then operates in absolute terms."

"Godspeed," Newton signs off, perhaps even adding: "I could think of a few experiments you may want to do. Also, since directionality can be changed in the real domain through collision, there ought to be a mechanism that changes directionality in the virtual domain."

Quantum Pythagoreans

Directionality is expressed as a form of polarization associated with moving real objects. In addition, it is not possible for the virtual object to change its own polarization autonomously. Such a change would require a computational exchange with another object. This would change the polarization of both entities in the framework of the conservation of the now-virtual momentum. The changed direction of movement becomes apparent once either object reenters the real domain. This operation can be visualized as a virtual collision because moving energy (energy and direction) is conserved while in its virtual form. For direction to be changed, virtual energy needs to be expended. Near-instant change in direction without any inertial acceleration is then possible, but only if the object (1) transitions to the virtual domain, (2) undergoes virtual collision, and (3) transitions back to the real domain. By transitioning to the virtual domain first, velocity (speed and direction) can be changed computationally and without g-forces. Staying in the real domain exclusively, inertia will always engage to conserve the real energy as well as the direction of a moving object, and the ship along with its occupants are subject to the familiar g-forces.

Vivibration is a word to describe virtual energy, which differs from the vibration of real things. A virtual entity does not have the inertia of a mass entity but does have vibration now called vivibration, which is equivalent to the moving energy of the real mass entity. Mathematically, the Planck constant **h** is a virtual number and thus de Broglie's equation relating momentum to vibration now relates momentum to vivibration.

If the classical physicist insists on describing the virtual electron in single particle terms then problems continue to pile on. Classically, a particle is something that is spatially well defined and moves along a path. Descriptions of the superposition pattern then becomes such that 'the path of the electron-as-particle is a path of all possible paths,' which sounds and is very similar to the intractable traveling salesman problem. Intractability arises when the disappearance and consequent wave behavior of a free moving electron is nonetheless modeled as a particle. In selected circumstances and when the interacting geometries are symmetrical structures, the 'path of all possible paths' becomes computable because the electron-as-particle maintains a computable probability distribution. Particular geometries of structures such as symmetrical gratings present the electron with *converging* geometries that create the environment for mathematical solutions. Electron optics or symmetrical physical structures *overlay* many paths the electron wave may take and such path reduction results in many paths that, nonetheless, have a bounded distribution. Paths then consist of many different overlaid paths that may result in a particular profile or a pattern. If the resulting pattern is computable, the appropriate description of such arrangement is computable overlay, tractable overlay, polynomial overlay, bounded superposition, or convergence. In the simplest example, if all cities the traveling salesman wants to visit are moved onto a straight line, the traveling salesman problem is no longer intractable. One can make an assertion here about the atomic structure, for atomic geometries present the electron with a symmetrical environment that allows the electron to have optical and computable interactions within the atom. Otherwise, electrons would leave the atom even without external energy enticement because the electrons would not find a computable environment.

Geometric structures can be generalized even further. The increase in organization is driven by computability. If the virtual entity is presented with a geometric structure that facilitates computability then the virtual entity always interacts with such structure. A structure, however, is made of real things. Virtual entities need real entities as particular structures to enable virtual entities to organize in a particular context. Real entities, on the other hand, need virtual entities' properties of nonlocality and superposition to resolve intractable problems with concurrent methods by engaging the virtual properties' infinite superposition.

Intractability is an area of many difficulties and promises. On the one hand, intractability keeps scrambled data scrambled but, as a concept, intractability also denies answers to many practical problems. Basic duality is at work here because the exclusive property of matter avails things to have repeatable and therefore predictable behavior. However, even with the joy you get from the predictability and consistency of a well-handling automobile, the best car cannot get any better on its own. To make the car better, a comparison and relationships of many variables need to be analyzed and such relationships are inclusive because they overlap. The overlap of variables can also be called superposition because relationships are not inherently exclusive. The number of interacting variables may initially be small but because each new variable may in general relate to every other variable, the total number of relationships raises rapidly; the number of all possible relationships **r** relates approximately to the number of variables **n** as $r = n^2$. Recalling the property of virtual variables, every variable in a relationship must interact with all values it acquired. At present there is no machine that could manage all variables carrying all values concurrently, which is exactly what electrons and photons do when they instantaneously and concurrently and as a whole interact with complex three-dimensional geometries of the environment.

Free electrons reduce completely when measured. The act of measurement forces the electron to supply a yes-or-no answer and subsequently the reduced electron is at a particular spot and nowhere else. However, if a free moving electron encounters one million non-uniform slits, the electron's wavefunction pocket instantly reshapes but does not reduce as it maintains its virtual computing environment. Any number of slits, then, presents the electron with an optical environment that is managed with concurrent mathematical computing. In general, today's computers can only emulate the unbounded environment sequentially and any absence of symmetry makes such problems intractable as well. When the detector reduces the electron's position to the *yes* answer the electron becomes computationally manageable on today's computer, but the concurrent mathematical interaction with multiple slits no longer happens because the localized electron cannot and does not interact with a plurality of slits in concurrent fashion.

An electron's path is modified when the virtual electron's wavefunction is modified. The positional probability of an electron's appearance corresponds to the electron's wavefunction and, therefore, any change to the wavefunction is a change in the electron's path. Because the electron wavefunction can be changed through computational means in up to 3D, an unusual interpretation and application opens up here: the electron's path can be changed with logical interactions.

Virtual And Irrational Numbers

A free moving electron becomes a virtual electron while carrying the dependent property of instant reduction, which is equivalent to instantly making its wavefunction shape smaller. There is, though, one additional aspect to the electron that was introduced by Dirac – that of the virtual positron. The virtual positron is attached to the electron and facilitates the mechanism of vivibrations. The virtual positron is the electron's agent of the conservation of energy. When the real electron spreads, it continues to be attached to the virtual positron. When the real electron spreads, it becomes the virtual electron that is now capable of manifesting the wavelength being held by the virtual positron. The virtual positron can be visualized as energy casting or a form, which the spreading electron mimics in the absence of physical interference or optical interactions that otherwise prevail and reduce the virtual electron into a real electron. Under all conditions the electron in its real state is attached to the virtual positron at a single point. Real and virtual entities are not computable unless they interact in but a point. Real and virtual entities are computable only through a process of reversible transformations.

The virtual domain is about entities that have opposites at each end. An electron's spreading parameter forms a pair of opposites because the electron becomes nonlocal in either or in both the positive or negative direction. The pair of opposites is then a single variable that is *double-ended*. The virtual positron is the keeper of the measure of the electron's energy while the virtual electron provides the measure of the electron's spatial spreading. A free electron, then, consists of two double-ended variables – two pairs of legs, so to speak.

Dirac encountered the virtual positron mathematically by applying virtual numbers. He thought of the virtual positron as a preexisting entity in space that joins the electron in some way. However, since energy must be added to break up and separate the electron and positron components, Dirac's view needs to be modified somewhat because the electron and the virtual positron are strongly linked at all times. A virtual positron carries the electron's energy in both the linear and rotational directions and thus the virtual positron contains vibration and spinning energies. A real electron's transitions to and from the virtual electron are shown below.

Quantum Pythagoreans

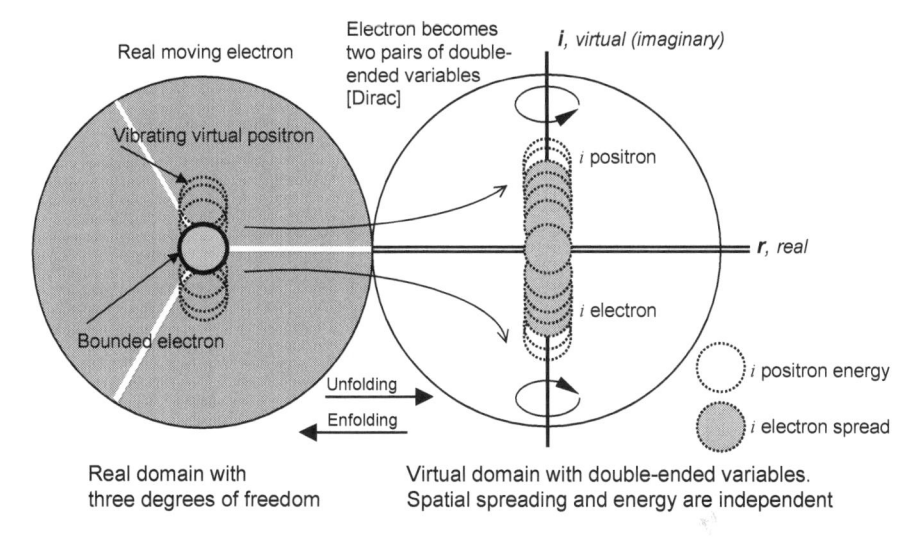

Real moving electron

Electron becomes two pairs of double-ended variables [Dirac]

i, virtual (imaginary)

Vibrating virtual positron

i positron

r, real

i electron

Bounded electron

Unfolding

Enfolding

i positron energy

i electron spread

Real domain with three degrees of freedom

Virtual domain with double-ended variables. Spatial spreading and energy are independent

Electron spreads with virtual energy of vibration and rotation

A good way of visualizing the electron after enfolding – which can also be called realization or reduction – is that the virtual positron surrounds the real electron while having vivibrations commensurate with the real electron's moving energy. The virtual positron then also enables the real electron to unfold, or spread, because the electron's energy is conserved during the transition of the real electron into the virtual electron. The act of enfolding converts the double-ended variable into a single-ended variable of a real entity, in this case the real electron. Dirac was also first to make a significant improvement in the explanation of the electron. Instead of taking the particle and the wave in a framework of complementarity,[38] he saw the electron as undergoing transformation from

---

[38] Niels Bohr initiated complementarity as a framework for describing wave-particle duality and possibly as a way to get discussions going. The basic difficulty with complementarity is that the real and virtual aspects are described in static terms because transformations are ignored. The transformation process, however, identifies parameters that change and the changed parameter's character becomes mutually exclusive with its pre-transformation character. Complementarity does not consider the either-or of parameters' character and complementarity is then stuck in "weirdness." The treatment of wave-particle duality through the transformation from one modality to another is more complete and is easier to comprehend: Some properties of a particle transform and, upon becoming wave-like, behave consistently and in character after the transformation. Without

123

one state into another, and back again. Electron transformations can then be discussed as the acts of unfolding and enfolding. Logically, moreover, electron and positron pairing in the virtual domain indicates that the increase of either end of the opposites is independent – that is, while the two pairs are centered about zero at the origin the spreading of the virtual electron is independent of the spreading of the virtual positron. Another way of visualizing it is that both double-ended variables are joined at the origin while either end of the opposites is free to increase or decrease independent of each other. The virtual positron may be changing its energy value depending on certain circumstances while the virtual electron could be changing its spatial reach value that is dependent on another set of circumstances. Overall, the electron-positron pairing is stable as one entity and this arrangement will also be amplified in the upcoming chapters as space-energy continuum.

The electron-positron pair becomes two virtual pairs in the virtual domain and the pairs' virtual aspect is defined by the *i*-operator – that is, *i*positron and *i*electron are two virtual pairs of opposites that computationally interact in the virtual domain. Any virtual variable can be aligned about its zero while each end of the opposite can modify its value independently. The (presently) horizontal axis, then, should be thought of as a backbone, bridge, axis, channel, or a projector to and from the real domain. The axis provides alignment and centering for all virtual variables – that is, the zero value of all virtual variables is on the horizontal axis.

The concept of the duality of the real and the virtual has been around for a while. Duality has also appeared in such forms as 'orderly and creative,' 'formal and informal,' 'fixed and volatile,' and 'lion and eagle.' It is now worthwhile to look at the virtual aspects mathematically but

---

differentiating the parameters that transform from those parameters that remain invariant, the particle-wave complementarity resembles the fine art of obfuscation. To take advantage of quantum mechanics for the purpose of travel and self-organization, the transformations of interest are those that are reversible. Also, a particle transforms because it has a moving energy that is momentum. The reversible transformation is between momentum and a wave and so the proper way of saying it is 'momentum-wave duality.' Mass (particle) transformation into energy is not reversible and will be discussed further on.

Quantum Pythagoreans

without losing the tie-in to physics or to the everyday use of virtual variables.

Mathematically, virtual aspects are described with *i*, which is a square root of minus one. *i* is not a real or an irrational number because no such number multiplied by self becomes negative. However, *i* is not just defined for convenience because computational results of many practical problems yield a square root of a negative number. A square root of a negative number, however, becomes positive if a square root of minus one is defined as a new unitary operator that operates not by multiplication but by *factoring*. The expression √-4 can be advanced and becomes √-1√4 and, subsequently, *i*2. √-1, or *i*, is not a number but it is a factoring, or extracting, operator. Operator *i* is in the family of other operators such as those for addition (+) or multiplication (·). Operators in general need one or more numbers or variables to operate on. Geometrically, the operation with *i* results in the rotation by 90 degrees counterclockwise and the number that is operated on enters the virtual domain. Drawing a right angle ⌐ as if opening a door is another way of visualizing the rotation operation. A similar way of "opening door to the virtual domain" is with the ⌐ symbol.

An operation with *i* results in a virtual number. *i* is shorthand for that operation and when the operation has taken place the *i* is attached to the number. *i* indicates that the number or a variable is now virtual, in the same vein as number -a where the minus sign is attached to indicate a negative real number after (-1)·a = -a operation. The minus sign, just like *i*, indicates and identifies the operator and the result of the operation. *i* needs to stay attached because any subsequent operations will treat the number as a virtual number. Virtual numbers also have addition and multiplication as valid operations among virtual numbers. Multiplication, however, may result in transformation of the virtual number.

Factoring out √-1 is an operation that is indicated by italicized letter *i*. Unfortunately, most people think of the *i* as a number. However, an operator cannot be treated as a number. Take the equation *i* = 1/*i*. When both sides are squared the equation is correct. But when both sides are multiplied by *i*, the equation is incorrect. Even veteran mathematicians stumble over this equation until they pick up on the difference between a number and the operator. It is not possible to say, "Multiplication is **1** over multiplication" because we are dealing with the operator. It is possible to

125

say, "Division is **1** over multiplication" but only because division is *defined* as the reciprocal of multiplication. In the current case, the operator *i* is *sometimes* (after squaring) a reciprocal of itself because minus one is the reciprocal of itself. Squaring enfolds a double-ended variable into a single-ended variable and consequently all negative values "fold in" or "fan in" over and onto the positive values.[39] Symbolically, squaring would be expressed with .. a square □.

Physically, subjecting real numbers in the real domain to the construction of a square root of negative distance does not make sense. Analytically, however, the operation by *i* or with *i* can be visualized as transformation between real and virtual domains. *i* is the *transformation* operator. The idea is that in the virtual domain the parameter of spatial distance is *i*distance, and *i*distance does not dominate because the wave nature of all virtual entities is inherently nonlocal. Another way of looking at distance in the virtual domain is that the parameter of distance is not enforceable in the virtual domain. Both light and virtual electrons, for example, propagate and move in space but the parameter of spatial distance that virtual entities have is always a derived and therefore a dependent parameter that always follows and, therefore, always self-adjusts. Overall, it is an application of the earlier classification of the independent and dependent variables. For example, upon entering another medium, light instantly changes its speed and the length span of its wavelength instantly follows. Similarly, an electron instantly reduces upon measurement

---

[39] Yet another closure can be found here: Multiplying a variable by **-1** facilitates half-circle rotation about a point (point is at origin) for all *odd* functions. For all *even* functions the multiplication by **-1** facilitates rotation (or reflection) about the vertical axis. [Metaphysically, **-1** is not a unisex operator.] You can now make a good case that the **-1** multiplier is also an operator that has different outcomes for odd and even functions. Mathematically, $f(-x) = -f(x)$ for all odd functions while $f(-x) = f(x)$ for all even functions. Treating **-1** as the operator also resolves the long standing unease about **-1**. In the equation $-1/1 = 1/-1$ the question to resolve is: "How could smaller over greater be the same as greater over smaller?" Using the **-1** as operator resolves the question; the multiplying *operator* **-1** has the same result for numerator and denominator. For those who wish to pursue the operator in more depth, think of operators as forces. For those who wish to pursue the operator in more breadth, think of operators as feminine.

regardless of the distance the electron reached during its spreading because in the virtual domain distance must accommodate, and therefore must follow, other parameters. Virtual numbers represent virtual entities and, consequently, $i$distance is spatial distance that is now a subordinated – that is dependent, parameter. The spatial distance parameter becomes the dependent parameter that *instantly* adjusts, and thus the act of superposition among virtual entities has no associated time parameter. In the real domain, a distance is a real parameter that is dealt with by supplying energy and thus moving a real thing a particular distance. In the virtual domain $i$distance changes its value by itself because no additional energy is needed. The idea of the virtual number goes back at least a millennium, but the interpretation of virtual numbers is lagging because our qualitative understanding of the independent and dependent variables – as well as its change at transformation – is deficient.

The instantaneous nature of superposition, moreover, adds to the Aristotelian notion of infinity. Aristotle treated infinity in a logical and formal manner, and derived the existence of infinity as something that exists but also as something that is "in potentia." It can be said that Aristotle's notion of infinity is that of the *unbounded*, which happens for all real things in the real domain. Aristotelian infinity was improved upon with the mathematical concept of *limit* where the values that compute the function are increasing and are technically reaching infinity, but the value of a function itself converges to a finite and bounded result. The virtual domain, however, contains infinity with two characteristics. First, the act of superposition is applicable to an infinite number of virtual entities because their wavefunctions instantly superpose without regard to the *quantity* of the superposing entities. The inclusiveness in the virtual domain, then, holds for an infinite quantity of wavefunctions. Second, the instantaneous nature of superposition discards *time* as a parameter of the new infinity. The real domain has infinity that is the classical, Aristotelian, in-potentia infinity. In the virtual domain, however, infinity happens that can be called true infinity. The first and the most important application of true infinity deals with tractability, a property that is the focus of this book from the very beginning. Any real system that is intractable will not happen in nature, for nature operates in real-time and not in need-more-time. The only way to make an unbounded and intractable system into a tractable system is in the virtual domain because the superposition there is instantaneous and there is also no limit on the quantity of superposing entities.

While the infinity of superposing entities may come together temporarily in any volume of space when photons are focused for the purpose of image formation, for example, infinite superposition also holds for wavefunctions that may come together and stay together during computational interactions. Because the superposition is instantaneous, the act of superposition is not the limiting factor in light speed propagation. The operation of superposition is the generalized form of the operation of addition and subtraction. The infinity in the virtual domain deals with any and all virtual variables without regard to quantity or time.

Pure mathematicians call virtual numbers imaginary numbers. Complex numbers have real and imaginary components but the virtual domain does not contain any real elements. This may seem to contradict the mathematical representation of complex numbers, which are generally written as **a** + *i***b**, where **a** is a real number and *i***b** is a virtual number. Carl Gauss popularized mathematical complex numbers notation as a summation of both the real and the virtual components. Real and virtual entities share the same space and they are always ready to interact – but they do not interact through addition and thus remain separate as **a** + *i***b**. Summation works – and can be mathematically advanced – for all virtual entities because virtual entities always superpose, but real entities' superposition is limited to identical objects. It is not possible to add apples and oranges unless these become identical as pieces of fruit. Fruits and vegetables cannot be added unless these become items of food. Real entities can continue with superposition only with ever-increasing generalization that ends with the top category of *energy*, which is common to all things. This then sets up the logic for the law of the conservation of energy because energy is the largest common denominator there is. Superposition is the general form of addition but the only way both the real and virtual entities can superpose is when real entities transition into the virtual domain. Should such transformation be reversible, and it can be, so much the better. Any and all transformations happen in the context of energy conservation. A domain is a collection of particular characteristics,

Quantum Pythagoreans

and space does not form an insurmountable partition between the real and virtual domains.

Pythagoreans originally did not apply the zero-dimensional dot of the Tetractys the way it is done in today's arithmetic. For the purpose of geometry, Euclid around 300 BC defined a point as "that which has no parts." Historically, zero presents difficulties to its users and zero's powers, or the absence thereof, is at the center of it all.[40] Zero – whatever it is – is acceptable in both the real and in the virtual domain even though in the real domain zero is nothing (not-a-thing), while in the virtual domain zero is infinity. It is understandable why ancient Greek mathematicians did not fancy zero. Most of us do not find any utility for no ships, no stadiums, and no wine. In the virtual domain the variables are double-ended because they vary on both the positive and the negative end, and zero is then at the center of the positive and negative values of all virtual variables. Zero or zeroing in the virtual domain is then also about centering. There is infinity of variables in the virtual domain and there is then infinity of zeroes in the virtual domain. Values of all real variables, however, start at near zero and vary in the positive direction – that is, all real variables are single-ended.[41]

Zero is not required in the real domain but its inclusion allows visualization of the real object's disappearance. Pythagoreans thought of zero as the absence of things, and zero then separates objects that can be visualized as blank, emptiness, nothingness, or vacancy. Numerical zero, as it evolved in India around 600 AD, started as a dot before acquiring its present empty oval shape while serving as a placeholder for numerals when using positional notation. Mayan zero is eye shaped and is drawn with many slight variations, but inherently it has the same placeholder function.

---

[40] Al Khowarizmi compiled the Indian numerals 0 through 9 that became known in the West as the Arabic numerals. Initially copied without zero, the number zero was in time included. Presently, the Indian numerals are called the Hindu-Arabic numerals but they are clearly Indian in origin and evolution.

[41] Mathematically, squaring of the wavefunction has two effects: The i is attached to a virtual variable and squaring reduces i variable into a real variable while the double-ended property of the pair of opposites becomes the single-ended property of a real variable. Squaring, then, transforms a virtual entity into a real entity.

● ● ● ○ ○ ○ ○   **The Virtual Domain**

A blink of an eye is perhaps the closest Mayans came to appreciating instant action. To some, the eye has close association with knowledge and the infinite. In the Mayan tradition, zero is an empty clamshell with lentil shape that is a placeholder for numbers and for people as well.

Pythagorean science of numbers consists of two basic pursuits: Numbers are about magnitude and multitude. It is apparent Pythagoreans assigned these qualitative properties of numbers to the real and the virtual variables, respectively. Pythagoreans did not treat virtual variables in the mathematical framework of the square root of minus one. Yet, Pythagorean teaching always addressed the pairs of opposites as something that is significant as well as commonplace. Even today the concept of the pair of opposites is best explained in economic terms of income and expense, while the square root of minus one is nowhere to be found. The variable in question is *earnings*, which is double-ended and has income and expense at each end. Income and expense vary in size on their own while earnings is a centering point, which is the net value of income and expense over a particular period. The variation in the value at both ends, and the net value, happen as the variable interacts with its environment. Over a particular period the variable acquires many values and this is the *multitude* (of values) that is unique to virtual variables. Every variable that has supporting and detracting qualities and quantities is a virtual variable. In English, most virtual variables have the prefix *in_*. Words such as intangible, inconspicuous, indispensable, and incoherent all represent virtual entities that acquire a multitude of values, and each have a continuous range between two extremes.

Single-ended real numbers, however, are about the representation of real things. Measurement of real things results in *magnitude* at a particular instance of time that yields useful results. A real variable has but one pertinent value of, say, "It's too hot in here." A virtual entity consists of many values – a multitude of values – and the useful result involves the collecting and processing all values from a particular period because the virtual variable changes during such period by interacting with other variables through a relationship. Discovery and the *measure of relationships* through the multitude of values is the needed useful result issuing from changes in virtual variables.

Quantum Pythagoreans

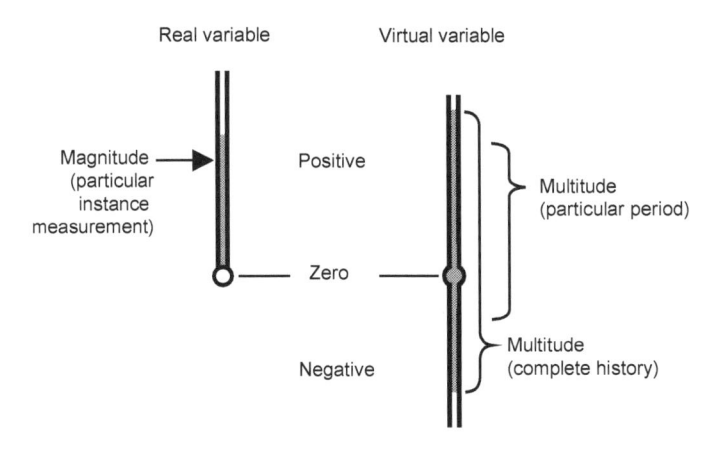

Numerical values of real and virtual variables

Real numbers represent real variables while virtual numbers represent virtual variables. Of interest to note is that, while virtual numbers are pairs of opposites, the opposites are not fundamentally in opposition. The virtual variable pair of opposites have extremes at each end (such as income and expense), but these opposites are not in conflict with each other because they are the high and low directions of a single, though double-ended, variable (that is earnings). Describing the virtual variable as a double-ended variable could be more appropriate than calling it a pair of opposites. The entire collection of values that is the multitude of numbers in a particular period defines the virtual variables' character. A single value for one sale and one expense does not give much meaning to the variable 'earnings,' and the context of a *period* necessarily comes along with every virtual variable.

Symbolically, double-ended virtual variables are expressed as double-headed entities. Consider, for example, the double-headed serpent Ometecuhtli of Mesoamerica and double-headed eagles of Europe. In the same category is the four-legged or two-paired Sphinx of ancient Egypt and we will engage the Sphinx for just that reason in future chapters.

In addition, real and virtual variables exist in separate domains but real and virtual variables are also not inherently in opposition with each other because they are qualitatively different. Also, much has been said about dualities such as hot and cold or wet and dry but the first thing to look at is that both of these variables are measures of real entities that are single-ended variables and the high and low is but a subjective offset from

131

the absolute zero of single-ended variables. A virtual variable is double-ended and has no limit at either end while a real variable is single-ended and has no limit at but one end. A real variable increases or decreases and at any particular instance has but one value. A virtual variable is not only double-ended but, in addition, either end changes independently of the other. Both ends of the virtual variable increase and decrease independently and the current value of the virtual variable is composed of all values between the two ends acquired during a particular period of interaction.

A real variable needs only one value of a particular magnitude to commence a process and to cause things to happen. Virtual variables are about relationships that are influences, and their interactions results in a priority among virtual variables. To quantify influence, however, a single measurement will not do and a multitude of a variable's values are necessary which, in turn, necessitates a compilation of values over a particular period.

In summary, single-ended real variables and double-ended virtual variables are not in opposition with each other. Also, values that real variables may acquire are not in opposition with each other, and values that virtual variables may acquire are not in opposition with each other. The only entities that are in opposition are matter and antimatter because, mathematically speaking, these entities are not computable when they interact and in consequence annihilate each other.[42]

Schrödinger derived the equation that models electrons diffusing around the atomic core, and the way he derived it illustrates the meaning of i nicely. He took a point charge representing the core and put electrons into the mix using complex variables – that is, variables that have real and virtual components. The inclusion of the virtual i-based component yields the Schrödinger equation. Without i the equation results in a classical solution where the electron-as-particle radiates energy as it accelerates around the core and energy conservation diminishes electron speed until

---

[42] Real and virtual entities are not computable except at one point that is the zero-dimensional point. There will be more to say on antimatter in the upcoming chapters.

the electron runs into the core. Schrödinger's equation takes the perspective of an entity that behaves as a wave. If the electron takes the shape of a closed wave with some degree of simplifying symmetry then Schrödinger's equation yields solutions that are easy to visualize. Matrix operations, on the other hand, take the perspective of generalized states of energy in 3D, and the shape of the electron could then be more general while accommodating electron reduction (enfolding) as well. Solutions of some matrices called Hermitian consist of only real components and these are called eigenstates. Eigenstates are spatial points where the electron can become a real electron. Eigenstate results from a full reduction of the electron's wavefunction where the reduction is the end point of a transition from the virtual to the real domain. To appreciate the eigenstate is to visualize the diffused (virtual, unfolded) electron reaching many spots around the core but there are only a limited number of spots where the suddenly real electron can appear, albeit temporarily, with real mass and angular velocity.[43] Eigenstates happen because the conservation of energy

---

[43] Real-virtual transformations are mathematically obtained with matrix multiplication. In the computational process of obtaining eigenstates, multiplication of an unbounded square matrix by a particular single-column matrix produces another single-column matrix that differs by a constant. Because both single-column matrices differ from each other by a simple multiplier that is the same in all rows, the energies before and after transformation are equivalent and subject only to a scalar multiplier between the two energies. Real and virtual energies are thus equivalent and mathematically related through a multiplier. The conservation of energy can be classified as a universal process that stems from a multiplication by a particular matrix. If certain relationships hold even for unbounded square matrices then the creation of universal constants could be explained. Matrix mathematics is also related to incomposite (prime) numbers when matrices are populated with random complex numbers bounded by Gaussian distribution and their conjugates reflected about the diagonal. Such mathematical environment is equivalent to the atomic environment because the atom can be subject to various photonic energies (represented by imaginary numbers) impinging on various electron energies (represented by real values of the electron reducing into eigenstates). Incomposite numbers live up to their name. A particular sequence of numbers, including larger prime numbers, do not ratio into a circle (orbital) and thus interfere with orbital mechanics. Lack of particular orbital segments then result in "forbidden" electron orbitals. Yet incomposite numbers construct all other numbers through their mutual multiplication and together they also construct (compose) the states of the "allowed" electron

holds at all times; the electron with a particular virtual energy value can materialize only with the same real energy that corresponds to a particular classical velocity and particular distance from the core. Because energy is conserved at all times, the wavefunction reduction is instantaneous.

In one's mind the virtual entity may invoke a unit that is spatially well defined through its wavefunction but, in addition, the virtual entity remains whole and is thus enclosed in one volume of space. In its interaction in the virtual domain the virtual electron can be computationally pierced into a shape of a doughnut but it cannot be partitioned into a plurality of discrete and spatially separate volumes.[44] One particular closure of the electron dual-slit experiment now becomes apparent in that the electron must spread into a large enough spatial volume to envelop the slits as well as the thickness of the material used for the slits. The virtual electron then moves through as a cloud of spatially bound frequencies while always consisting of one – that is, contiguous – spatial volume. If at any time the virtual electron cannot remain within one spatial volume then a portion or the whole of the electron's wavefunction instantly reduces for the purpose of retaining one spatial volume. The appearance of the superposition pattern then allows gauging of the depth of the virtual electron because it is the same as the thickness of the material used for the slits.

Another question concerning the real electron becoming the virtual electron is the status of the parameter of the electron charge. While it is certain that the charge is conserved because the reducing electron continues to have it, the virtual electron's charge spreads along with the electron and continues to exist because a free moving – that is virtual – electron can be steered or accelerated by an electromagnetic field. Any electron, real or

---

orbitals. Calling prime numbers *incomposite* numbers is very close to their atomic function.

[44] It appears that theoretical math is, at times, removed from its physics basis. Theoretical math treats two-dimensional topology as having discontinuous boundaries between the top and bottom of the object while the real world two-dimensional object such as a sheet of paper, pancake, or spiral galaxy have sharp yet continuous boundaries at the edge.

Quantum Pythagoreans

virtual, continues to interact with electromagnetic fields. Electrical charge, then, is an invariant property during transformation. However, while an *accelerating* charge, such as the charge of the electron, always radiates an electromagnetic field, the atomic electron does not radiate. The absence of radiation from the atomic electron is not due to some qualitative transformation of the charge as such but it is due to the geometry of the charge. An atomic electron always spreads in a geometry that is radially symmetrical with respect to the atomic core. Radially symmetrical position of any portions of the electron cloud does not produce accelerating movement of the electron *as a whole* and, since the charge as a whole is not accelerating, the electron does not radiate an electromagnetic field. As long as the electron is symmetrical about a point, the electron will not radiate energy, for it is spread in the orbital around the nucleus. Classically, an electron must radiate if it is modeled as a point charge moving around the core. In actuality, as a point-symmetrical standing wave the electron's charge is spread and not moving. During orbital changes, one can visualize the atomic electron as radially advancing, retreating or changing into altogether different radially symmetrical geometric shapes such as a sphere, doughnut, bowtie, teardrops, or flower petals.[45] Radial symmetry is symmetry about a point. If any part of the electron is spinning then the corresponding radially symmetrical part of the electron would be spinning in the opposite direction. Because the electron always consists of contiguous volume, the symmetrically counter-rotating electron could pass through the center of the core but at very low probability density. For some orbitals such as the doughnut, the geometry of the electron is radial but it does not have a counter-rotating component that a bowtie shaped orbital could have. Some orbitals, then, do have a net angular momentum. De Broglie found a relationship between the electron's linear momentum and vibration frequency but he did not work on the angular momentum. It can

---

[45] A change in radially symmetrical shape is also quantized because the atomic electron wavefunction has a continuous boundary. As the wavefunction wraps around the centrally located core it can smoothly close on itself only at discrete intervals (Newton's *fit* or the ancient Egyptian *closure*) that are in increments of its wavelength. Radially symmetrical shape is quantized in that the possible shapes have discrete increments in spatial volume. Because the change in wavefunction is instantaneous the change in volume shape or a "jump" is also instantaneous.

● ● ● ○ ○ ○ ○ The Virtual Domain

be shown that the angular momentum becomes the rotational frequency. The updated virtual electron, then, is a contiguous volume of space containing vibration and rotation frequencies. De Broglie did not identify the frequencies as virtual frequencies but because they are virtual frequencies they are best called vivibrations and it is vivibrations that computationally interact with the environment.

Earlier in the chapter, real numbers were shown to have real magnitude while virtual numbers make sense only as multitude. The irrational numbers need their own category, which is separate from either the real or virtual numbers. Irrational numbers cannot and do not have a magnitude because magnitude has a definite and finite length. The irrational number, then, exists as two points in space or as two zero-dimensional points. The irrational number cannot exist as a one-dimensional entity. Irrational numbers are about *separation* between two nodes. Separation exists as distance between two nodes but the separation is not filled in with a one-dimensional solid entity because the distance would then acquire real magnitude. Therefore, irrational numbers are about open and unique separation between two zero-dimensional points. The number of wavelengths that can exist between the two points spanning irrational distance is infinite.

Putting numbers on a number line attempts to simplify visualization of all numbers, but the only way of doing so with irrational numbers is with zero-dimensional end points (nodes). The number line is not a solid line and neither is it a line of magnitudes. The number line is a direction of one-dimensional freedom and is best visualized as a dashed line. The irrational number sees the number line as an empty centerline separating two nodes.

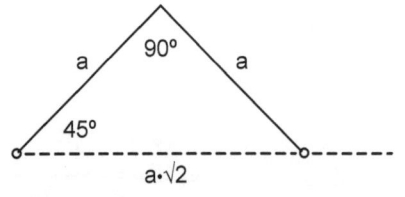

Irrational number: separation between two zero-dimensional points — $a \cdot \sqrt{2}$

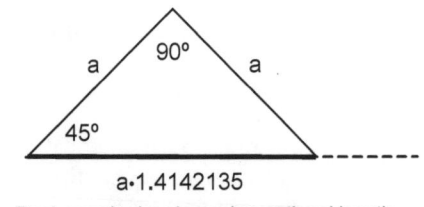

Real magnitude: always has rational length — $a \cdot 1.4142135$

Irrational and rational construction of square diagonal on a centerline or 'number line'

Quantum Pythagoreans

Squares can be constructed using exact and real angles of 45 and 90 degrees. Other irrational numbers can be constructed as separation between two points but the two points are not guaranteed to fall on one common centerline or number line – that is, the general construction of any and all irrational distances cannot overlap one centerline at will. Only the specific subset of irrationals, that is, diagonals of any and all squares, can be constructed to overlay any particular centerline. If an irrational point does not fall on a given centerline, the point cannot be moved there because the act of rotation or translation necessitates the taking of a magnitude, which irreversibly converts the irrational number into a real number.

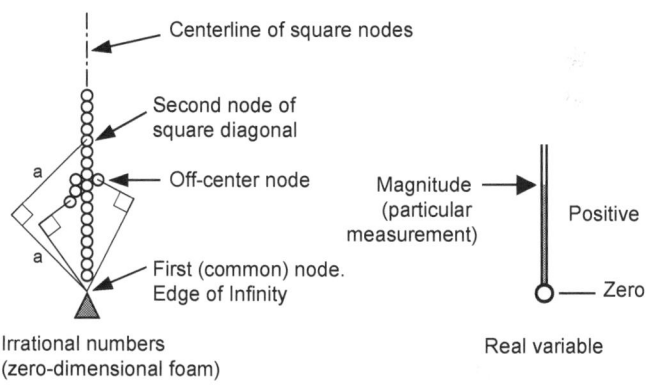

Numerical values of irrational and rational (real) variables

Fundamentally, the construction of irrational numbers is only through the Pythagorean theorem, which, in turn, applies and relates spatial distances in two dimensions. The construction that results in two points also spans the distance of the diagonal (hypotenuse). One of the two points is referenced in one location that becomes the Edge of Infinity or Infinities. This point is the starting point of all constructions. The infinity label indicates that the span of irrational separation accommodates an infinity of wavelengths for each irrational number. The Edge of Infinities will also be the anchor for an infinity of irrational square diagonals. Moreover, the diagonals of all squares are the only irrational numbers that can be placed

137

on a given centerline.[46] If each of the second nodal points is visualized as a small circle, the allusion to foam is apparent. We'll see how the placement of irrational nodes on one straight line can be applied inside the Great Pyramid. See the Appendix.

The virtual domain is about concurrent relationships among the intangible and infinite quantity of variables. Real parameters that are measurable in the real domain become, after transformation, computational parameters in the virtual domain. Looking forward to upcoming chapters, one area will be about the independent variables in the real domain becoming the dependent variables in the virtual domain – and vice versa. In the virtual domain the concurrent interaction among the infinite number of variables results in a new and qualitatively different computational modality.

Complete Model of the Free Electron

To make the electron's model complete, we need to describe the interactions of the real and the virtual electron with representative entities and we need to determine the priority of such interactions. Because the electron is well evolved, it is expected that that priorities will be well established and repeatable. An atomic electron is computationally bounded within the atomic structure and there may be mathematical representations of the atomic electron in other literature. In general, electron evolution inside the atom is bounded when the electron forms a wave closure, which is also a solution to the Schrödinger equation. Wave closure is a form of a standing wave. Standing waves exist in linear geometry while orbital wave closures exist in spherical geometry. The electron inside the atom is not localized and, while the atomic electron is not freely evolving without bounds, the electron has a family of discrete orbital solutions.

A free electron can be bounded or unbounded. It becomes the virtual electron when it starts evolving (spreading, diffusing, unfolding)

---

[46] Stated without proof. However, the centerline itself can first be established by an irrational number other than a square, which is indeed the case in the Great Pyramid.

into space. A fully bounded (fully reduced) electron is a localized electron and is, therefore, a real electron. Light binds a free electron and such an electron is partially or fully reduced to the extent of the light's wavelength. Light's wavelength prevails over electron's spatial evolution and, for example, the superposition pattern disappears.

Light also actively interacts with atomic electrons when light enters matter. Although light's energy remains unchanged, the interaction changes light in other ways. Light's path changes through interaction with orbital electrons called refraction, and light speed also slows down during the interaction. Light polarity may also change during the electron bounding process. This can be visualized as a small turn in a photon's polarization that happens in some materials each time a photon binds an electron. Shorter photonic wavelengths bind the electron tighter still and greater interaction between photon and electron results in greater refraction. Higher energy photons having shorter wavelength will refract to a greater extent than lower energy photons. In orderly materials such as crystal prisms, the rainbow pattern nicely shows the refraction and consequent photonic separation based on photonic energy. Light interacts with free electrons in similar fashion. If light passes through a cloud of free electrons that have varying degrees of spreading, image distortion or "warping" would be observed. At object's edge there is usually greater accumulation of free electrons and image distortion may be particularly pronounced there.

Vivibration is the electron's energy. Light will interact with an electron's vivibrations during the interaction but only in temporary superposition. Electron vivibrations and photon wavelength[47] that are the respective net energies remain invariant in their mutual encounter. Light can manifest its energy only when it is absorbed and no absorption takes place in the interaction with a free electron. There is no energy transfer during interaction between light and a free electron. Refraction is optical interaction and no energy transfer takes place during optical interaction.

---

[47] In the upcoming chapters on light there will be a differentiation between photonic wavelength and photonic undulations. Photonic wavelength is the measure of photonic energy but photonic undulation is a measure of photonic presence.

● ● ● ○ ○ ○ ○ **The Virtual Domain**

Electron transformation between its real and virtual states is accompanied by the transformation of the electron's parameters as summarized below. A parameter that is or becomes the virtual parameter has a check mark in the shape of the letter **v** to indicate its virtual nature. The check mark also graphically illustrates that the virtual variable is able to unfold – spread its wings so to speak – and become a double-ended variable. Attaching *i* to each variable would do just as well but at this point **v** makes for a better graphical match.

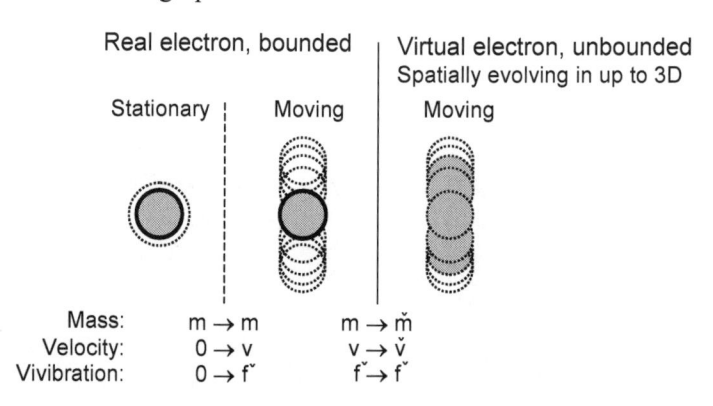

Real electron, bounded | Virtual electron, unbounded
Spatially evolving in up to 3D

Stationary | Moving | Moving

Mass: $m \to m$ $m \to \check{m}$
Velocity: $0 \to v$ $v \to \check{v}$
Vivibration: $0 \to f^{\check{}}$ $f^{\check{}} \to f^{\check{}}$

Transformation of electron parameters

Electron reduction also happens in an attempt to partition the virtual electron into a plurality of discrete volumes. Free electrons partially or fully reduce in all interactions with real matter. Moving real matter always has vivibration. Energy transfer between electrons and matter happens through their mutual vivibrations, which starts at zero and is potentially unbounded. Matter-electron interaction also raises the question of what constitutes a "collision." The corresponding energy transfer is a complex topic that requires much experimentation. All real variables are single-ended and real matter can transfer its real energy only in the direction of its velocity but this is not necessarily the case during the interaction with a free electron. Vivibration is energy in virtual form and a free electron vivibration can have zero value when the electron is stationary, or a positive or a negative value. Virtual energy can have positive or negative value because any virtual variable is double-ended. After the transformation from virtual to real energy, however, real energy cannot be negative and virtual energy is then transformed to a particular

Quantum Pythagoreans

forward or reverse velocity corresponding to the polarity of the virtual energy. Negative virtual energy, then, refers to direction rather than to some bad, detrimental energy. Virtual energy can also be transformed into localized real vibration that is heat. Real energy can only be positive because real energy is a single-ended variable; there is no such thing as negative work because work is produced by applying energy and work is created by applying energy going in forward, reverse, or any direction. The magical aspect is that electron's vivibrations can have its energy expressed with forward (positive) or reverse (negative) value and such electron interaction ("collision") with another object can slow down or speed up or deflect such object in any direction. The electron, then, can have vectored 3D vivibrations even though the electron as a whole is not moving. A "collision" is the energy transfer where vivibrations of the electron are transferred to the object and in the case of virtual energy the direction of movement of the object after "collision" can be in any direction because the directionality of the energy is inherent in the vivibrations. The complete representation of all of electron's parameters is shown below.

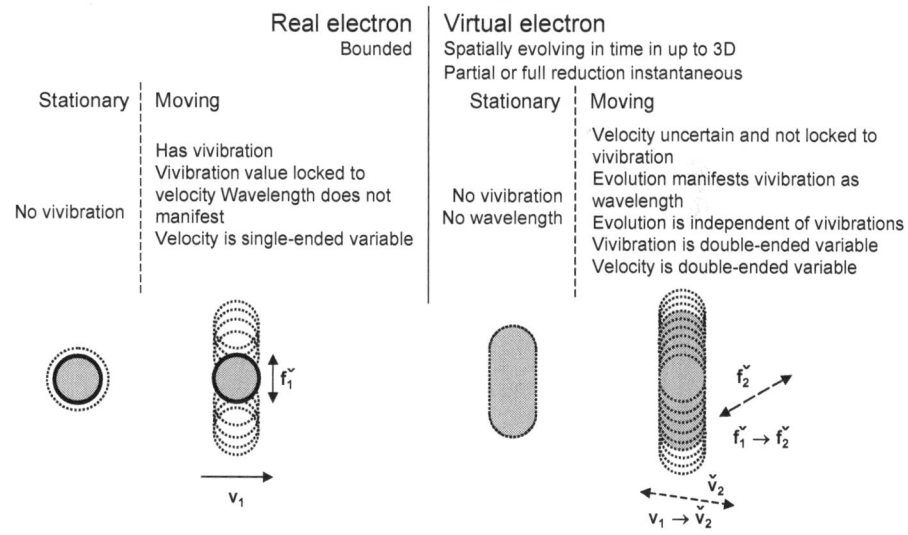

	Real electron Bounded		Virtual electron Spatially evolving in time in up to 3D Partial or full reduction instantaneous	
Stationary	Moving		Stationary	Moving
No vivibration	Has vivibration Vivibration value locked to velocity Wavelength does not manifest Velocity is single-ended variable		No vivibration No wavelength	Velocity uncertain and not locked to vivibration Evolution manifests vivibration as wavelength Evolution is independent of vivibrations Vivibration is double-ended variable Velocity is double-ended variable

Electron energy: Transformation of velocity and frequency parameters

Perhaps the most difficult situation to reconcile is when a bounded moving electron has vivibration but does not manifest it with a wavelength. How could a moving electron have the frequency of vivibration but not a

141

corresponding wave? The moving bounded electron is localized and its vivibration is kinetic energy that is stored (or saved or memorized) for the purpose of energy conservation. Vivibration results from the mechanism of energy conservation that manifests as inertia. When an electron remains bounded the electron cannot concurrently interact with an array of atoms in the same manner as a wave – but the electron's moving energy is conserved *locally* nonetheless and it retains and conserves its moving energy through vivibration. Bounded electrons can no longer pass through both slits simultaneously in the dual-slit experiment because the electron's mass is localized.[48]

Inertia arbitrates the conservation of energy and direction. If a body speeds up, inertia requires a certain amount of energy to be expanded and a certain amount of work performed. The very same amount of energy is stored locally with the moving object in the form of vivibration, and this holds for any real object. The work expanded to speed up an electron is stored by the virtual positron and includes the energy as well as the direction in which the electron was accelerated.

Stationary free electrons are virtual electrons that have no vivibration. They evolve within all three spatial dimensions as per Schrödinger's equation and their spatial presence does not impede the movement of real matter and does not refract light. While it is true that light will partially or fully reduce free electrons in their mutual interaction, the reduction is temporary and electrons evolve as soon as light passes on. Electron spreading happens in the absence of interactions with photons or comparable physical structures, and is potentially unbounded. Free, evolving, and stationary electrons are ether. Ether is composed of virtual electrons that do not have vivibration and, therefore, do not contain organized energy. Free and stationary electrons that have no vivibration

---

[48] In 1927 Clinton Davisson and Lester Germer used the crystal of nickel for the electron target and confirmed the moving electron's wave property. The same experiment can be performed with light illuminating and thus bounding the moving electrons. The wave property of the electron will then cease to manifest but the value of the electron's energy impinging on the crystal will not change. If Davisson and Germer took this extra step they would have been well on their way to discovering the mechanism of energy conservation.

Quantum Pythagoreans

can be called ether or *uncommitted* electrons or unprogrammed, neutral, or voided electrons. Uncommitted electrons of ether interact with matter and light through reduction, can be captured in a structure to make matter such as hydrogen, and can be imparted with organized energy through logical computational means. One can also say that movement is one way of adding energy and thus adding a measure of organization to the electron. Uncommitted electrons do not facilitate movement of other organized entities in an active way but they can become the building units of such entities. Finally, electrons are not subject or a party to gravitation and do not gravitationally interact among themselves or with neighboring matter. Technically, however, electrons can be energized and vivibrated to interact with the atomic core and neutralize gravitation by some interfering mechanism.

It is most likely the ancient Egyptians called ether the Nun, the primordial stuff that is dark in all directions. Atum is the intelligence that uses Nun to commence the creation of materia and consequent separation of the sky (heaven) Nut and the earth Geb.[49] The separation between the sky and the earth is important and Shu facilitates it, for Shu is a whooshing sound of the diagonal conduit that separates and bridges these disparate domains. In more recent cosmology such as found in the Old Testament, ether is referred to as "water in the dark," while Shu is omitted altogether as if the heaven and the earth were just two things – instead of being the separate *and* joined domains they are. Shu is not and cannot be a straight conduit but is a transformative agency between the virtual (heaven) and the real (earth) domains. Transformations between the right and the left brain in the human happen "by itself" as most of us take it for granted, but in the construction of a machine the process of domain transformation needs to

---

[49] Atum is said to either spit out or ejaculate his heaven and earth creations. The creation, then, appears to be a logical operation. After the idea is conceived and worked through, the idea needs to be released, let go, or consciously sent out. The logical system continues to hold together as one embodiment, not because of physical proximity but because of logical likeness. The new creation must then be detached or ejected in some fashion if it is to leave its creator and possibly manifest as new creation.

be explicitly addressed. If Shu is omitted, the isolation between the two domains is analogous to a lobotomy.

Many writers equate the primordial darkness with chaos, but uncommitted electrons contain no organized or disorganized energy. The darkness, in its original meaning, refers to the pre-sun environment and also to the environment that has no vivibrations and, therefore, as the environment that has no energy content. In the ancient Egyptian cosmology of the Heliopolis region, Atum is associated with a growing mound and, through spin (rolling ball), gives rise to the sun. In play on words, the atom (Atum) was made from something that is none (Nun).

The existence of spread out and uncommitted electrons of ether does not leverage the science and philosophy of everyday life. The existence of ether, then, may not rise beyond the question of belief. The speed of light, however, is fixed and absolute because of ether. In the past, the absolute lightspeed has been taken as a given from a measurement and without further explanation. Moreover, once energy is imparted onto ether and the electrons of ether are no longer uncommitted, everything changes in the most profound and existential way.

First Conjunction. The Atom

After J. J. Thomson discovered the electron in William Crookes' evacuated glass tube, Thomson promoted a model of the atom as a 'positively charged diffused core embedded with separated particle electrons.' While the Crookes tube was becoming the X-ray tube, the television set, and colorful neon lights, the Thomson model was not advancing very well. From experiments and further work by other physicists, the theory that later prevailed was that the atom looked more like a 'positively charged and clustered particle core surrounded by diffused electrons.' Yet, coming back to Thomson's model, and even with apparently incorrectly reversed attributes, Thomson's model is simpler and technically more compelling because it does not need to deal with localized charges repelling each other. In fact, the mechanism holding the core's charged components together is not understood beyond a backfill presumption that there ought to be some new, strong force that overrides the repulsive core charge, but only for short distances. Scientists often bring up the criteria of Occam's razor, which is supposed to favor those explanations and theories that are simpler and therefore cut closer to the truth. Science writers, however, are not kind to Thomson's model of the atom – but for some reason cannot ignore it. Thomson's model is called the "plums in pudding" model at times, the "raisins in pudding" model at other times, the "raisins in bread" model in another variation, the "watermelon with seeds" model in a different cultural context, and the "cranberry cake" model in yet another new-fangled incarnation. Thomson, however, not only quantified the electron's properties but he was also the first to propose the radical notion that the electron is a component of the atom, which up to that time was supposed to have been indivisible. Thompson's diffusing idea was also the first; the core's components were not in one location but were spread in space. Thompson's charge diffusion was radical and ahead of Schrödinger's mathematical treatment of the spreading and wave-becoming electron, and ahead of Heisenberg's position uncertainty concept that allows the particle to spread. So, in light of Thomson's validated foresight there is merit to exploring how much proof there is in Thomson's pudding.

It is most surprising that, since the outreach of the atomic electron is well understood, Thomson's model of the atom and its core has not been updated with the atomic particle spreading or with a "becoming a cloud" mechanism. In spite of past misgivings, the electron nonlocality inside the atom is well established.[50] There is no good reason for continuing to hold on to some kind of extra strong nuclear force for close to eighty years if a working and simpler mechanism is available. Furthermore, the presumed

---

[50] Electron diffusion is presently computed and modeled with much fidelity. Bohr, however, refused to accept that the atomic electron's position becomes nonlocal and the only way for him to keep up with the point-electron view was to ignore the fact that atomic electrons do not continuously emit photons – in defiance of the linear or circular radio antenna where electrons always radiate energy as long as they are being accelerated. When the image of the electron is limited to a localized entity, it is most difficult to conceive of electron spreading in the explanation of the dual slit experiment, or in the explanation of the Davisson-Germer experiment in which the angle of electron incidence is not the same as the angle of departure.

145

strong nuclear force is first postulated, but then it could be bypassed with atomic wavefunctions that "tunnel" right through the force in order to explain alpha particle emissions. Particle tunneling is synonymous with the spreading of the virtual particle via its wavefunction and it is the subsequent reduction of the nonlocal wavefunction that accounts for tunneling.[51]

The present model of the atomic core is static but the real nature of the atomic core is dynamic. The atom is composed of a positively charged and pulsing core surrounded by diffused electrons. The pulsing of the core results from the core's components periodically spreading and instantly reducing as real particles. The core's "beating heart" components can and do instantly reduce and localize. The basic difference between the core's components and electrons is that the electron's localization is event-driven and happens only when particular photons enter or leave the atom, while core components localize periodically. The core as well as the electrons can and do spread to interactively and concurrently compute in the virtual domain. So far as the theorist can surmise, the protons and neutrons on the inside – and electrons in the orbitals on the outside – maintain some computational and therefore dynamic balance for the purpose of

---

[51] Alpha particle is the core of the helium atom. Alpha particle's reducing wavefunction allows some quantity of alpha particles to appear outside of the core with subsequent ejection from the atom.

In 1895 Wilhelm Röntgen discovered X-rays, which are high-energy photons. In 1896 Henri Becquerel affected photographic film with luminescent salts when the film was otherwise shielded from light. It turned out the luminescent salts were also radioactive salts. Becquerel was able to seize the moment and show that such radiation – now called alpha particle radiation – differed from X-rays. 1897 saw the discovery of the electron and in that same year Santiago Cajal's work culminated with Charles Sherrington naming the synapse, a unique structure interconnecting brain cells. Planck's quantum impulse constant introduced in 1900 kicked off the 20th Century fast and furious. In 1902 Philipp Lenard experimentally discovered that light's ability to release electrons from metal – the photoelectric effect – was a function of light's frequency (photon's energy) rather than light's intensity (photon quantity). Also in 1900, posthumously for Gregor Mendel and thirty-four years after his publication, three biologists proclaimed that Mendel's quantization of inherited features into genes made sense – with dominant and recessive gene attributes making sense as well.

perpetuating the atom's existence. Upon interaction with external photonic stimuli the electrons can reduce individually, albeit temporarily, in one of the real nodes offered by eigenstates. Proton and neutron diffusion inside the core has a significantly shorter reach than that of the electron because a proton's mass is considerably larger. What comes to the forefront, however, is that core components of protons and neutrons operate in the virtual domain as well – that is, a proton's positive charge periodically becomes a diffused positive charge and interacts with an electrons' diffused negative charge.

The mechanics of the transition of the atomic core from the real to the virtual domain is through its radial movement, for as soon the core materializes it starts to expand from its center through charge repulsion, but the velocity of the expanding core components gives components the wavelength they need for nonlocal interaction. The expansion of core components then becomes spatially bounded. In the simpler case of the hydrogen atom the single proton in the core begins to diffuse in accord with the Schrödinger equation. In most, if not all, cases the electron's wavefunction passes right through the center of the core while also retaining a radially symmetrical distribution about the center of the core. The resulting blending and geometric distribution of positive and negative charge clouds then accounts for atomic stability – that is, atomic components maintain a computable relationship despite external disturbances.

The orbital behavior of electrons is understood to a much greater extent than the behavior of the core's protons and neutrons. In 1885, Johann Balmer found out about four unique wavelengths of light radiating out of the hydrogen atom. Suspecting a relationship between these four wavelengths he succeeded in deriving the empirical formula that yielded these four numbers with accuracy of four or more significant digits: $\lambda = b \cdot m^2/(m^2-n^2)$. Number **b** is a constant and contains the Planck constant **h**, which was not proposed for another fifteen years. Numbers **m** and **n** are integers while **m** is always greater than **n**. Squared relationship of numerator **m** hints at orbital geometry, which also means that number π would be inside the number **b**. In reverse, Balmer may have thought of orbital geometry early on, and from among many who tried, he applied and succeeded with square relationships.

147

Balmer's integers **m** and **n** are the numbers of the principal atomic orbitals. Orbitals then exist in jumps where **m** and **n** are the starting and the ending orbitals, or vice versa. Balmer's equation relates two principal atomic orbitals and quantifies the wavelength between *any two* principal orbitals **m** and **n**. The key role of integers also surfaces the idea of quanta because wavelengths corresponding to non-integers do not exist. Today, each principal orbital carries a number representing the principal or main quantum number that behaves according to Balmer's equation. Subsequently discovered additional atomic numbers are but refinements of the principal quantum number. The "next" quantum number deals with the elliptical shape (eccentricity) of the principal orbital while yet another refinement deals with spin direction of the electron that changes the principal orbital a bit. In a great and total Pythagorean recall, integers commenced their rule in the atom. The possibility of understanding invisible atoms is simply too big a prize to keep locked up in a barn. Balmer's equation opened the doors wide and there was no way of going back. The world has been subject to a multidirectional stampede ever since, nobody knowing whether the blitz could ever be contained.

Balmer's equation reveals that the reciprocal of wavelength – that is, energy – is proportional to a difference between two squared quantities **m** and **n**.[52] The four wavelengths that Balmer's equation matched correspond to **m** = 3, 4, 5, and 6 while **n** = 2. Boldly and correctly Balmer proposed that other wavelengths would be found corresponding to combinations with **m** = 7 and **n** = 1. In time, the quantity **n** = 1 was equated with the lowest, also called "ground," state of the atomic orbital. Balmer's equation also confirms that several orbital levels can be skipped during an electron's single orbital jump because photons exist that correspond to **m** = 4 or **5** while **n** remains at **2**, for example. Twenty years later Niels Bohr derived a similar formula incorporating the Planck constant, but his formulation uses only the number **n** as the principal quantum number. Bohr's equation is difficult to understand because it is

---

[52] Some authors rework Balmer's equation as $f = 1/\lambda = R \cdot ((1/n^2)-(1/m^2))$, which emphasizes frequency and has a format similar to the harmonics series.

Quantum Pythagoreans

not apparent that an energy differential can span more than one level.[53] Using both **m** and **n** indicates **m** and **n** are free to acquire different integer values. Balmer's equation nicely shows that electron jumps are not confined to the nearest two orbital levels and this becomes significant when in future chapters we will equate the harmonics series **1/n** with the ancient Egyptians' fractions. Balmer's original presentation in the form $\mathbf{m^2\text{-}n^2}$ hints at a Pythagorean relationship, which carries better visualization when considering the proportions and the function of the King's Chamber inside the Great Pyramid. See the Appendix.

The transition from the real to the virtual domain is accompanied by a transformation of some of the real particle's attributes or parameters. Pythagoreans found it engaging that certain operations between numbers always resulted in a change to the number's even-odd parity. Pythagoreans established distinct sets of operations: one set resulted in even-odd transformations while other sets of operations kept the even and odd numbers invariant.[54] In the mid 19th Century, Evariste Galois developed what he called the group theory. It took another fifty years before grouping – such as the even-odd number grouping of Pythagoreans – caught the attention of other mathematicians once again. What appeared to be an unexciting set of transformation and invariance rules evolved into a mathematical group theory and this was applied to atomic physics during the second half the 20th Century. Today, all atomic components belong to either the even or the odd function group. Photons belong to the even function group. The interaction of a photon with the electron in a gas molecule can transform some of the electron's parameters while the balance of parameters are invariant. Mathematically, the interaction and the

---

[53] Bohr did not go further in the explanation of the orbital atomic separation in hydrogen except by declaring that the orbitals between Balmer's principal orbitals were, well, forbidden. Some insight can be made through the application of Riemann math and his search for the role of incomposite (prime) numbers in the geometric context of spheres. Incomposite numbers can well provide the rationale for orbital separation.

[54] In his *Elements*, Euclid uses even and odd properties of numbers to prove that the square root of two is an incommensurable (irrational) number.

possible transformation of an entity's parameters is modeled by the act of *operation*.

A particle directed at matter finds the atomic core materialized or diffused. The core can rebound the particle or the core can retain the particle, and if the imparted momentum onto the core is high enough the core can break up. Real momentum can be mathematically equated with heat energy and core break-up temperatures are computed at billions of degrees. However, once the core becomes the virtual core, another virtual particle can interact with the core computationally and then a mathematical solution, rather than just heat or momentum or pressure, yields the answer that results in the particle's fusion with the core. It is likely that certain computational interactions can build just about any core of the element from the periodic table, while other previously unknown elements can be created as well. It is but stable computational solutions within the core that create different elements of matter.

Fast forwarding to upcoming chapters, gravitational interaction among atoms materializes the core every time the gravitational wavefunction reduces and imparts momentum to the core where the momentum manifests as force of gravitational attraction. The atomic core can then be visualized as oscillating between its real and virtual states where the reduction into the real state of the core is accompanied by momentum creation that manifests as gravitation. The atomic core, then, has an eigenstate having periodicity in its solution while the constant pulsing behavior of gravitation can be likened to the master synchronization of a digital system. Indeed, Newton's gravitational constant that relates gravitational force to mass and distance has time as one of its units of measure.

It is also conceivable that matter can become suspended – that is, in the absence of the gravitational force the atomic core can remain in its virtual state and matter's mass parameter disappears. Gravitation would need to be neutralized first and the eye of the galaxy looks just right for that, for in the eye the gravitational force from the entire galaxy sums up to zero. It may seem that the eye of the galaxy is densely compacted with matter, but the absence of gravitational force means that matter remains in its virtual state. The computational nature of gravitation, however, makes gravity neutralization possible without actually traveling into the eye of the galaxy.

The Virtual Nature of Light

Light has unique properties. It is easy to show that light is a virtual entity despite the presently prevailing notion that light is a real entity. This can be traced directly to an erroneous presumption that photons of light carry real momentum and, consequently, light ought to be able to push a mirror by bouncing from it. Yet nothing can be further from the truth. In the early 20th Century the idea of energy-as-inertia was debated and the mathematical equations were manufactured to show that 'matter has inertia and if matter is convertible to energy then energy has inertia.' The key to manipulating equations into agreement with such logic was to insist that photons of light have real properties because even though mass may transform to energy the inertia does not. Inertia, however, is the property of mass and inertia is the derivative of mass. Inertia is *defined* as force that needs to be supplied if mass is to increase its speed. Energy must be, by definition, imparted onto an object if an object is to increase its speed. Photons of light, however, readily decrease and increase their speed when moving in and out of matter such as glass but no energy needs to be imparted to photons for them to speed up or slow down.

If the photon of light is presumed to be a real entity with "effective" mass, then photons of light could be treated as miniature balls of mass that exert pressure when rebounding – and light's mass balls could then be attracted by gravity as well. However, neither light nor light's path is subject to gravity. The myth that light can be captured by gravitation results in belief in black holes that is only a belief because it does not rest on facts. The actual measurement of a laser beam's zero pressure at the mirror surface is the direct proof that light has no inertia. Light has no momentum during reflection and the logic of 'matter has inertia and if matter is convertible to energy then energy has inertia' is clearly erroneous. It is a presumptuous conclusion that when matter transforms to energy the inertia of such matter remains invariant.

151

● ● ● ○ ○ ○ ○   **The Virtual Domain**

For real matter, inertia is engaged to uphold the conservation of energy when energy is used to speed up real matter. Light does not need inertia because the energy of each photon is fixed for the life of the photon. Light cannot be accelerated or decelerated by applying energy to it, and the conservation of energy holds by dealing with the photon as a particular quantity of energy where the quantity of energy is based on a photon's frequency.[55] A particular photon of light cannot be injected with additional energy and energy cannot be extracted from a photon other than by the absorption of the entire photon.

In retrospect, light was used as needed but without consistency. When light photons are presumed to be bouncing balls, no explanation is offered on how a photon can rebound at one time but be absorbed at another time. Selective absorption of photons by gas has no classical explanation. Light cannot interact classically as bouncing balls if certain wavelengths are absorbed and, as the wavelength shortens, photons go right through – yet, continue to shorten the wavelength and photons become absorbed once again by the same gas. Equations are but shorthand for logic. People may disagree with certain postulates while making hypotheses of their own. Mathematical equations can be just as corrupted as someone's logic. Laboratory and field tests go a long way toward resolving conflicting mechanisms but the most useful mechanism is to think in the framework of commercial applications. For example, the claim of the Compton effect is that a photon scatters free electrons, but since both the photon and the free electron behave as waves, such a proposition is tenuous to some – and unworkable to others. Tests were performed claiming a confirmation but while a photon can impart energy to the *atomic* electron at absorption, a photon cannot impart some or all of its energy to a *free* electron. The atomic electron shares the photonic energy with the core and may leave the orbital in the act of ionization, but a free electron cannot be scattered with a photon because the energy of a photon cannot be transformed into real moving energy without the framework of momentum

---

[55] Energy of a photon depends on the frequency of electrons' orbitals that produce the photon. There will be more on photon's frequency in the chapter on the Planck constant.

conservation. Laboratory tests were performed on atomic electrons but the results were erroneously claimed to be just as good for free electrons. Compton received the Nobel Prize for the discovery of photonic scattering of free electrons. However, no commercial device has ever been built based on photons that scatter free electrons. The Compton effect is a defect that got past the review committee, possibly in the excitement over the quantum nature of light. Most likely, however, Compton received the Nobel Prize because light's presumed real momentum was in line with theories of those doing the selection. The absence of physical pressure from a laser beam at reflection directly invalidates the Compton effect proposition.

In the arena of academia the equations are handed down for safekeeping. There is resistance to changing them until something breaks. It is the hard-earned knowledge and publication of such equations that makes them difficult to change, particularly if such equations are in government proposals or if projects the likes of solar sailing are underway. Light, however, cannot be treated classically even theoretically. Lightspeed measurement accuracy, for example, is improved by sending a crest of light over a larger distance. For a particular distance the number of bounces – be it one or one thousand – has no effect on the measurement of lightspeed. Light rebounding does not slow light down, so light then cannot be modeled as a spring that gets squeezed during rebounding. Inertia is defined as a force that resists a change in velocity of an entity, in this case the photon. Yet if photon speed were to slow down to zero during rebounding under the presumption of inertia then the slowing down and speeding up would delay light's propagation. On the other hand, no object carrying inertia can instantly reverse direction without slowing down first.

Should the presumption of light's momentum be accepted, a perpetual motion machine could be built, as follows: The photon is directed at a mirror. After the first bounce the mirror should have received some energy and start moving but the photon that imparted such energy is moving in the opposite direction at the same speed as before the bounce and without losing any of its own energy. A second mirror is placed parallel to the first, and now the photon imparts energy to the second mirror and then heads at full speed back toward the first mirror. Under the presumption that photon has "effective mass," a photon has the same energy after the bounce as before the bounce. After the first bounce there

153

exists excess or over-unity energy and even if the photon is absorbed instead of bouncing the second time, there exists unaccountable excess energy. The presumption that the photon of light carries momentum is untenable. There simply is no classical or other model that would give light inertia and transfer light's momentum during reflection – except with corrupted equations.

Professional groups could acknowledge that the absence of light's pressure at reflection invalidates large portions of today's physics and astrophysics. Once the truth is out about the absence of light's pressure, also called radiation pressure, then certain courses could not be taught without the obvious linkage to fraud. Charging fees for courses for which the subject is known to be false is fraud, pure and simple. Presently, the charge of fraud is avoided by feigning ignorance about the absence of light's radiation pressure. The hope, then, is that some higher institution of government will appreciate the net benefits of pursuing the facts, even though some professors would need to be retired or retrained. Another way to resolve these issues is with a grass roots movement, resulting in the perception that paying for certain education or services is an investment without a future. In the worst case, entire cultures and populations just wither away or are overcome by technologically more astute groups while the historians swiftly, glibly, and with the benefit of hindsight, dismiss defeated cultures as carrying excess baggage.

In the arena of public opinion, people readily state a preference. Some people embrace the fact that the electron can become an invisible and vibrating packet of space-energy continuum. Others prefer that only stage magicians be given a license to make things disappear. Some people become so smitten with the idea of a laser beam pushing a payload out to space that they simply do not want to perform the experiment that would invalidate such a possibility. Yet others see light's nonlocality as the unfolding of the destiny that up to now was denied to humans. Powerful computational and logical tools – verbalized, visualized or symbolic – will help you resolve very complex problems and discover the most unusual solutions. In the energy-as-inertia case, real matter has inertia but the transformation of moving matter's energy into vivibration conserves the energy while the parameter of inertia – as well as mass, velocity, and position – are no longer needed for energy conservation and become subject to transformation.

Equations that relate mass to energy are not reversible transformations. Reversible transformations transform momentum into vivibrations but do not transform mass into energy. Once mass converts to energy it ceases to exist as matter, and matter cannot be rebuilt because the transformation is for all practical purposes irreversible.[56] All transformations of interest here are between various forms of energy that – be they real or virtual energies – are reversible. While light's transformation to energy is, strictly speaking, not reversible because the absorbed photon is gone, new photons with identical features can be readily created when photons are absorbed within finite and computable structures. Absorbed virtual entities such as photons will become part of a structure if photons find closure there. Photons that are absorbed within the computable structure of an atom or molecule add energy to the structure, which manifests as increased volume or pressure. The photonic

---

[56] Relation $\mathbf{E} = \mathbf{m \cdot c}^2$ describes the conversion of matter into energy but matter subject to this conversion cannot be recovered because the matter-to-energy relationship is not directly reversible. The equal sign in the equation, then, does not properly describe this relationship since this is a one-way process with a left-pointing arrow, and the conversion is actually $\mathbf{E} \leftarrow \mathbf{m \cdot c}^2$. Consequently, light's mass cannot be obtained from the one-sided relationship if placing a photon's energy for **E,** and this also corrects algebra's weakness of not recognizing irreversible transformations. Even though the conversion happens in the framework of the conservation of energy and a photon is produced, it is most difficult to convert such photon back into matter because X-rays and gamma rays are hard to work with. Almost all X-rays and gamma rays leave a solar system that undergoes a nova or supernova explosion and much of the real matter is lost during such explosion. Some science writers make a claim that, in a laboratory, the application of energy "creates" both matter and antimatter and tout it as the proof of the creation of matter from energy. However, it is the destruction of matter that is being described because the application of energy first breaks matter down into a matter and antimatter pair. Subsequently, matter and antimatter converts to energy as per the one-way (irreversible) process. Science writers making a claim that energy creates matter and antimatter ignore the fact that the energy needed to create matter and antimatter is much less than the energy arising out of the subsequent matter and antimatter annihilation. Matter and antimatter is addressed in future chapters.

transformation is then reversible because photons can in time be recreated through emission.

Symbolically, a quantum of uncommitted and propagating energy finds closure as Ouroboros, which is similar to Newtonian 'fit' or ancient Egyptian 'closure.' There are two kinds of Ouroboros, but the one with greater fidelity is the one that *bites* a piece of its tail rather than the one "eating" its tail.

Unbounded

Light is electromagnetic radiation consisting of individual photons of light that can be visualized as wave crests of particular height and length but of very little thickness. Light does not influence other light or other light's path because light is inclusive without bound. In one locality many and all wave crests of light can meet, overlay each other, and continue on a straight path without affecting each other. Each photon of light, then, is independent of all other photons. However, each photon of light changes its path instantly by local interaction with matter. A photon's interaction with matter modifies a photon's wavefunction but photons do not permanently modify each other's wavefunction. Local superposition among photons does not leave any logical imprint on any photon – that is, there is no memory or priority associated with light's superposition that temporarily arises among individual photons. The overall effect is that photons just pass through each other.

When light is not absorbed in its interaction with matter then light has an *optical relationship* with matter, and light retains its virtual properties. Reflection and refraction are two examples of optical interactions that were described over two thousand years ago.

The slow acceptance of light as a virtual entity is a combination of psychological and cultural bias, which consists of not questioning the current mainstream theories as long as some expectations are fulfilled. Various and often contradictory theories of light have made their appearance since antiquity and it is not always easy to consider all proposals. However, to forgo a simple experiment that would validate or invalidate the pressure a laser imparts on a mirror points to clear absence of scientific leadership. Avoiding for forty years a laser experiment that provides many answers also puts to rest a boast that science is inherently self-correcting. Light does not and cannot convey pressure at reflection and books written on solar light pushing sails through space, equations and all, are but chronicles of empty promises of reality that cannot deliver. Volumes written on ships deriving thrust from ejected or reflected X-rays or microwaves are about fraudulent efforts to proceed with expenditures without simple validation. Momentum cannot arise in any optical interaction.

Momentum Arises

Another barrier in accepting light as virtual is the fact that we can see light and feel light as heat. Virtual things are, by definition, intangible. However, light can become real in the form of heat and electrical energy. Light interacts with and is absorbed by real matter when it lands on the eye's retina. When a photon makes the transition from virtual to real it becomes real energy and that is how light is recognized or measured. Once the photon of light makes the transition to the real domain and becomes real energy, light ceases to exist as light and that is the definition of light's absorption. To see or feel light, the photon of light must be absorbed. As you look around, the only photons you see or feel are those photons that are being absorbed. There is more to photons and electrons that will be discussed in the chapter *Light, Just Warming Up*.

When a photon of light is absorbed, the entire wavefunction of the photon reduces. This assertion can be stated in even stronger terms. Light becomes real energy if and only if light is absorbed, and light is always absorbed if momentum can be created in the context of momentum conservation.[57] If the photon is not absorbed in its interaction with matter, light's path instantly changes or the photon branches subject to the atomic geometries it encounters and this manifests as optical qualities of matter.

---

[57] Light's ability to create motion can be associated with Aristotle's *primum mobile*. Gravitation also results in movement and you may want to decide which mover is prime, if any. Is it, perhaps, that light creates expanding motion while gravitation creates contraction and rotation? A photon can also have the linear and the angular component.

The Virtual Domain

When a light detector actually measures or detects light, the absorption (reduction) of a photon has taken place. The absorption is no other than the instant and complete collapse of a photon's wavefunction. When photons interact with materials the likes of glass, water, gas, or the sun's corona, optical interactions result in refraction, reflection, or interconnected branching of photons, which can be also called multipath. The actual measurement of light ends a photon's existence as a virtual entity. New photons, however, are created during the electron's transfers to lower energy orbitals within the atom where higher energy levels have been previously facilitated by absorbed photons. Light is perceived as heat at longer wavelengths, yet the transfer mechanism of reduction from the virtual to the real domain is the same for all wavelengths of light. While light can and does make the transition to real energy, there is no such thing as a real photon.

Light's virtual energy becomes real energy that manifests as real momentum in case the electron is emitted from the atom by a photon. Since momentum can be created only when the total momentum is conserved, light cannot move an electron in an encounter with a single electron, and the physical ejection of electrons can happen only for atomic electrons where momentum is imparted on both the electron and the core. Because two real things are needed, the ejected electron comes from one of its eigenstates where it is real. The electron, however, is not residing at some eigenstate, as the electron is spread out within the particular atomic orbital. The photon binds the electron to its corresponding eigenstate before momentum can be realized at both the real electron and the core. The ejected electron becomes electrical energy while the recoil of the core becomes local vibration that manifests as heat. Pressure can also develop if the photon-absorbing material is not allowed to expand. The creation of electrical energy through the photoelectric effect, then, cannot have greater than 50% efficiency.[58] A photon's virtual energy becomes both heat and

---

[58] Relationship between energy E and light frequency f is $E = h \cdot f$ where h is the Planck constant. Electrons manifesting the photoelectric effect are ejected from the atom. As the photonic energy is partitioned 50-50 between the core and the electron in order for the conservation of momentum to hold, the electron is imparted with the energy of $½h \cdot f$. Should the electron leave the atom, the energy

electrical energy on the eye's retina. Light, however, remains a virtual entity during optical interactions such as reflection, and light cannot impart momentum to a mirror because momentum would not be conserved. A photon needs at least two real things to become absorbed and get the two things moving in the opposite direction while conserving momentum, and this prerequisite is not satisfied at the surface of the mirror.

Heat And Pressure Mechanics

Light reflecting from the atom cannot move the atom. Light that is absorbed by the atom cannot move the atom. Light refracted by the atom cannot move the atom. If heat or any energy is to be converted to moving energy, it must be in the context of momentum conservation. A single atom can absorb photons and store such energy as an internal increase in the orbital kinetic energy of its components, but the atom as a whole cannot and will not move. If the atom cannot store any more energy, it will radiate excess energy as light but it still will not and cannot move. If two atoms come in close proximity to each other as they do in gas molecules, then real momentum can happen because two bouncing atoms can and do conserve momentum – and this is the true quantum mechanical explanation of gas pressure. Photons span certain spatial distance, and lower energy photons, or heat photons, span a greater distance. A single photon is absorbed when it finds itself between two atoms and touches both atoms. Two atoms then bounce away in the opposite direction thus conserving momentum. It is also apparent that different gas molecules will absorb photons of different wavelength because atomic distances are different in different molecules. More complex gases such as methane, or mixtures of gasses such as air, absorb photons of several wavelengths, for different molecules have unique separations between atoms. Atomic separation, then, can be measured by the wavelength of photons molecules absorb. Atoms in a molecule absorbing a photon will bounce in the opposite direction, but because atoms in a molecule are subject to additional binding

---

of the free electron is below this amount. The electron is ejected away from the atomic core and must move in radial rather than the tangential direction.

forces the atoms jump into equilibrium where such new equilibrium corresponds to higher gas volume. The additional binding forces are no other than valence electron orbitals that span and bind two or more atoms in a molecule with their orbitals – valence orbitals define molecular formation. Valence electrons absorb photons and form a new computable equilibrium, one photon at a time.

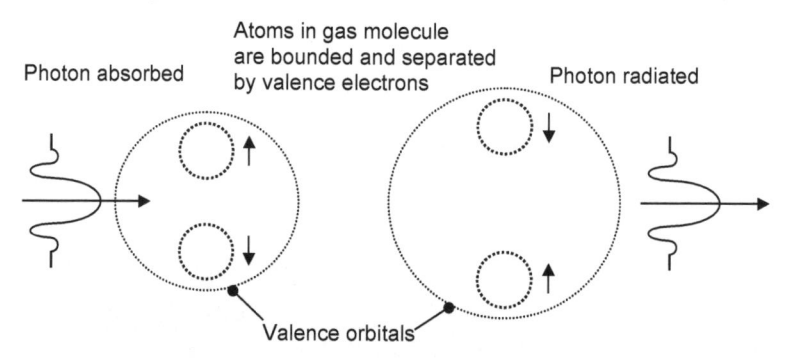

Molecule vibrates in place as photons are absorbed and radiated

Being a dynamic equilibrium, a gas molecule decreases its temperature and volume by emitting a photon and then the average atomic separation in a molecule decreases. A particular temperature of gas is thus defined as a dynamic balance where photon absorption and the corresponding out-bounce of atoms is equalized by radiation and the corresponding in-bounce of atoms within the molecule. A gas molecule continues to heat up only if additional photons are available for absorption. An instrument is then able to measure the temperature of a known gas by the wavelength the gas molecule absorbs or emits. When two atoms in a molecule are brought together by mechanical compression, a photon is produced which propagates out. Similarly, if a molecule of gas spontaneously emits a photon, two atoms bounce toward each other while both the gas pressure and temperature decrease as momentum is conserved. The mechanism here is the decrease in the orbital of valence electrons that is accompanied by a release of a photon. Mechanical compression of gas creates photons and if photons are reabsorbed within the gas volume, the temperature of the compressed gas also rises. In reverse, expanding gas readily absorbs photons from the environment, which results in the cooling of the environment.

Quantum Pythagoreans

This mechanism is not limited to molecules of gas. Photons are absorbed inside individual atoms as well but since the distances are smaller it is the shorter wavelengths of visible light that interact and are absorbed within the atom. The absorbed photon transfers one half of its energy to the core and the other half to the electron, which then moves to another, higher-energy orbital. As photons continue to be absorbed the electron moves to a higher and higher orbital until the electron gains enough additional energy to leave the atom altogether and the atom or the gas molecule becomes ionized.

As the incoming photons' wavelength becomes progressively longer the photon absorption chances diminish. Similarly, as photons' wavelength becomes shorter the electrons that can absorb such photons are those having their orbitals closer and closer to the core where the reach of the core provides orbital limit. One would then expect the absorption to peak at a particular wavelength and taper off on either side of the peak wavelength. If atoms can absorb a particular range of wavelengths then they must also radiate the same range of wavelengths. Different wavelengths of light radiated by objects can then be measured and indeed there is a peak in the emission of a particular wavelength, with decreasing emission on either side of the peak.

Presently, gas pressure is explained as molecules moving on a generally straight trajectory that bump into neighbors and walls after so many inches of travel thus creating pressure. Molecules are supposed to be kept moving by rogue billiard ball photons, which speed about while ignoring their absorption or the presumed energy transfer at reflection.

Gas molecules, however, vibrate *in place* as a result of absorption and emission, and it is the *localized vibration* of all molecules that manifests as pressure. One would expect the absence of incoming photons at absolute zero would also result in zero pressure and minimum volume. Quantum mechanically, gas molecules and atoms convert absorbed photons in expanding motion increments that results in volume increments while, in reverse, contracting decrements in volume are accompanied by photonic release. The dominant mechanism is that the molecule subject to a neighboring mechanical pulse will release a photon since the molecule cannot move far in a linear, or straight, direction. The behavior of a molecule of gas, then, rather than being subject to the rigidity of a billiard ball, is cushioned by a photon's release and re-absorption. Linear bouncing movement is the primary mechanism of sound conduction, as sound waves are periodic physical displacement of molecules. The heat conduction mechanism, however, is proportional to the average distance a photon travels before being reabsorbed. Since gas molecules in general do not have a preferred alignment, released photons travel in any direction. Statistical methods could then be used for determining the gradient and the net movement of photons that is commensurate with the net conduction of heat. The apparent opportunities in this area relate to directed and therefore extremely rapid conduction of heat, which could be as fast as the speed of sound. Crystals, for example, have a greater heat conduction rate than amorphous materials and this is likely because the symmetrical geometries create preferential photon radiation paths.

The volume of atoms and molecules increases with increasing temperature, and the ability of atoms and molecules to store photons in their internal structure is measured as *specific* heat, which is unique for various atoms and molecules. Specific heat, in turn, is a measure of the atom and the molecule to computably integrate the increasing amount of absorbed photons. Atoms and molecules are instrumental in converting the photonic form of energy into forms of energy that are readily reversible. Gasses and liquids subscribe to similar mechanisms. Brownian motion is a manifestation of locally vibrating molecules of liquid that come in proximity to small objects such as pollen.[59]

A physical barrier can be put in place such that no photons are allowed to enter or leave the gas volume and if such a system is completely isolated it then becomes a closed system. The idea behind the establishment of a thermally closed system is to provide a cylinder and a piston with a computable environment for the internal combustion engine. The computable environment, in turn, results in the formulation of the 2nd

---

[59] Brownian motion was originally explained as collisions with straight moving molecules. Molecules, however, vibrate in place and may transfer some of its kinetic energy to a neighboring particle. While both models are kinetic, the underlying mechanisms are quite disparate – one presumes linear motion while in reality it is oscillation.

Quantum Pythagoreans

law of thermodynamics, which allows statistical manipulation of a large number of molecules within a thermally closed space. Yet the 2nd law cannot possibly be applied to the universe because in the universe there are no barriers forming closed systems. The existence of a thermally closed system is the fundamental requirement for the second law to hold and this law then cannot be applied to the universe.[60] Moreover, no barriers are necessary if some photons continue to interact within the system while others just pass through it, having no affinity to the system and having perhaps a very small chance of being reduced within the system. Photons move about the structure by continual absorption and emission and then possibly leave the system if the environmental conditions change. Photons' absorption and emission also accounts for heat storage and heat conduction. Electrons, protons, and photons are capable of forming structures and systems without thermal or physical barriers. It is apparent that barriers are not necessary to form a system and a system can remain computable without barriers. It is also likely that particular photonic shapes that interact within physical structures we call atoms and molecules have

---

[60] There is much interest in drawing a line between science and pseudoscience. In the 2nd law of thermodynamics the thermal and physical boundaries establish a closed system, and these boundaries result in some new equations. When these equations are generalized to, say, the entire universe, the constructs that made these equations possible should then also be applicable to the entire universe – yet the universe has no thermal or physical barriers. Such generalization is an excellent candidate for the pseudoscientific mindset. Logically, it is taking the fish to the mountain but not taking the fish bowl along. Similarly, a 'field' is applied liberally to explain the phenomenon of gravitation. However, the formation of any field such as an electric field requires real structures such as parallel plates. When the pseudo scientist speaks of gravitational field, the field is presumed to exist even though the structure forming such field does not. Perhaps the best example of pseudoscience is general relativity that starts with an enclosure such as an elevator and then does not allow the experimenter to look outside. Subject to such constraint, the equations found in the closed elevator are then generalized to gravitation as a whole. Yet it is apparent that the construct of 'not being able to look outside' exists for the sole purpose of obtaining a particular equation. The pseudoscientific mindset tries to include artificial constraints as "natural" constraints; it adds constraints or presumptions because it cannot deal with the actual environment.

actually evolved to such shapes, and such shapes are then unique to atomic and molecular sustenance.

Light Mill

A light mill confounds almost all who go beyond enjoying the mill's rotating paddles. The explanation of this toy's movement eluded James Maxwell who mathematically formulated light itself.[61] A light mill usually has four paddles that have a bright and a dark surface on the opposing sides of each paddle. When illuminated, the dark side of one paddle as well as the bright side of another paddle is presented to light, each on opposite sides of the mill. Any difference in pressure resulting from the differences in the dark or bright character of the paddle's surface then allows the mill to rotate. A light mill is enclosed in a glass bulb while the inside is filled with inert gas such as nitrogen. The illuminated mill's paddles rotate in the direction such that the dark sides recede. The bright side, though, is supposed to receive twice the light's presumably real momentum because light's classical momentum is not only imparted but rebounded as well.

The dark side of a light mill paddle readily absorbs light of all frequencies and radiates energy at all frequencies, but the dark surface preferentially radiates frequencies having a particular distribution that favors longer photons corresponding to heat photons. At the dark surface a broad range of photonic frequencies is readily absorbed but mostly photonic frequencies of heat are radiated. The air molecules neighboring the dark surface absorb heat photons and two atoms comprising the gas

---

[61] James Maxwell's equations are describing a continuous production and propagation of electromagnetic radiation. Lightspeed can be calculated from Maxwell's equations and lightspeed value is independent of the velocity of the source that is producing it. Maxwell proposed an experiment that would send two components of light in two directions at once. In order to measure the speed of light relative to the speed of its source, one component of light would be sent along with earth's rotation while the other half would be sent in the perpendicular direction. The execution of Maxwell's proposal turned into Michelson and Morley experiment of 1887.

Quantum Pythagoreans

molecule bounce away from each other when momentum is created. The dark paddle surface thus receives some momentum from the expanding molecules, which is no other than increasing gas pressure. On the bright paddle surface heat radiation happens to a much smaller extent because bright paddles reflect almost all of the light, and air molecules are not exposed to the radiation of heat that gas molecules can absorb. A bright surface facilitates reflection of mostly visible wavelengths of light that go though the gas molecule without being absorbed. Also, a bright surface stays cooler than a dark surface because almost all light is reflected rather than absorbed. On the dark paddle's surface, however, higher temperature accounts for greater heat radiation and expansion of gas molecules with consequent increase in gas pressure. On the dark paddle surface the energy of a broad band of light transfers via heat to the increased real momentum and consequent increase in pressure of gas molecules that are in proximity of the dark surface. Light's energy results in the increased gas pressure over the entire dark surface and the dark surface moves and recedes from the light source.

By putting the light mill in the freezer, the dark paddles continue to emit and absorb heat. Gas inside the mill also emits heat as it cools down and the conservation of momentum requires atoms in the gas molecule to bounce toward each other with a consequent decrease in gas pressure. Since the dark surface of paddles is quicker to cool off, gas in proximity of dark paddles now preferentially radiates into the dark paddles because dark paddles are cooler than bright paddles. The radiation from gas is accompanied by the bouncing of two gas atoms toward each other, which is the only way to conserve momentum and where the total energy of the created inbound bounce inside the molecule equals the emitted photon's energy. The inbound bounce then accounts for decreasing gas pressure. One should expect a reversal in the direction of the light mill rotation when the mill is moved from a light source and into the freezer because gas pressure at the dark paddle surface then decreases at a faster rate compared to the bright paddle surface. Gas molecules at the dark paddle surface are radiating at a faster rate; the dark paddle surface acquires lower pressure and the paddles move toward lower pressure such that the dark surface advances.

● ● ● ○ ○ ○ ○ The Virtual Domain

The experimenter should also expect a continued gas evacuation around mill paddles to bring the light mill to a standstill and that is indeed the case. A light mill does not move when all the gas is gone. The classical physicist holds on to his equations, believing and proclaiming that the mill in a vacuum reverses direction because the bright side is classically supposed to receive twice the presumed momentum but such an experiment has never been performed. The classical scientist is quick to expound on a comet's tail pointing away from the sun as a proof of light pressure, yet the sun also emits quantities of real atomic particles that cause this effect.[62] When it comes to light, the classical scientist is at the end of his rope. Conflicted and mesmerized, the classical scientist is not only unwilling but also simply unable to direct a laser beam on a mirror to measure the presence or absence of pressure. The experiment measuring light pressure on a mirror puts all classical physicists on the spot and challenges them to rethink and rework the science of physics. The resistance by scientists to measure the presence or absence of pressure that a laser beam imparts on a mirror goes unabated. It is apparent that university scientists are a group held together by beliefs and fantasies rather than by the objective pursuit of physics and its applications. There is no doubt that the present science of physics is no other than state-supported religion, for the belief has become more important than the reality. The unwillingness to perform the experiment measuring photonic pressure on a mirror is an expression of the power of official belief. Attempts to use lasers, microwaves or solar light for propulsion – as well as making adjustment to mirrors with presumed laser beam pressure – are some of the present wishful thinking projects that are being funded without verification.

Separated But Undivided

---

[62] Newton drew many sketches of the comet in its orbit along with its tail but he did not hypothesize the mechanism that changes the tail's direction as the comet moves around the sun. Newton also found good partnership with Edmond Halley when he confirmed that Kepler's elliptical orbit geometry is applicable to comets as well. Halley accurately predicted a comet's return, which earned Halley's comet its name. Halley sponsored the publication of Newton's *Principia*.

Quantum Pythagoreans

When a particular manifestation spans distance but its validated mathematical portrayal does not include the parameter of distance then such effect is said to be nonlocal. The instantaneous reduction of the electron is a nonlocal phenomenon and light exhibits nonlocal behavior as well.

Inside the interferometer instrument photons of light encounter the half-silvered mirror. At one instance, each photon branches into a pair and each end of the pair moves at lightspeed in a different direction, say, in the left and the right branch.

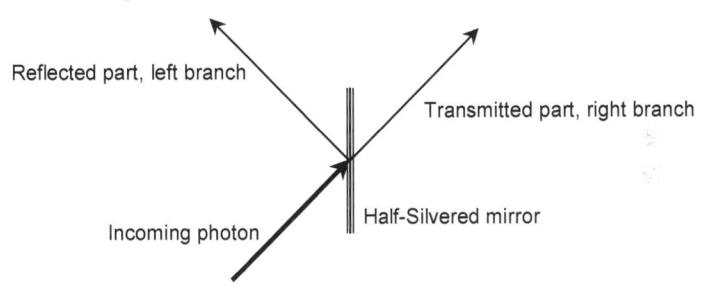

Reflected part, left branch
Transmitted part, right branch
Incoming photon
Half-Silvered mirror

Light beam splitter

The beam-branching mirror – at times called the beam splitter – branches photons at half the photon's virtual energy. Although each photon branches at 50-50, each photon remains whole and, therefore, both ends of the branched photon are interconnected. A photon of light can be branched and its ends separated but the photon cannot be divided.

After being branched at the beam splitter, the photon's wavefunction is spread with 50 percent of its virtual energy at each end, as shown in the illustration below. In one of the interferometer branches a detector attempting to reduce the photon receives a *no* photon reading in half of the readings and a *yes* photon reading in the other half of the readings. This is a heuristic result. The measuring instrument in one of the branches presents the photon wavefunction with the yes-or-no construct and receives a *no* photon answer 50% of the time. The measuring instrument interacts with the half of the photon every time in an attempt to reduce it and the *no* photon answer is just as good an answer as the *yes* photon answer.

The Virtual Domain

Photon wavefunction branched by half-silvered mirror

- Detector in left branch interacts with photon's wavefunction and receives no-photon reading 50% of the time
- Wavefunction at 50%
- Left branch | Right branch

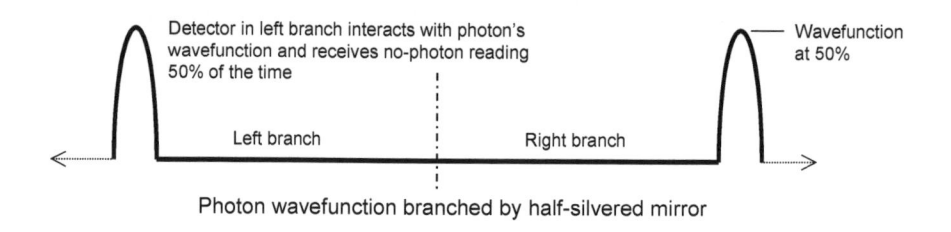

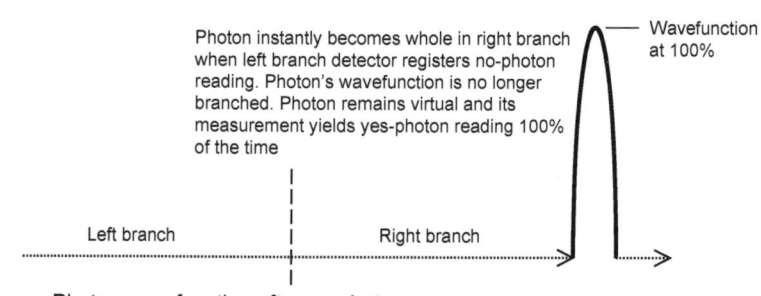

Photon wavefunction after no-photon measurement

- Photon instantly becomes whole in right branch when left branch detector registers no-photon reading. Photon's wavefunction is no longer branched. Photon remains virtual and its measurement yields yes-photon reading 100% of the time
- Wavefunction at 100%
- Left branch | Right branch

If the interaction with the left branch of the photon results in a *no* photon reading then the photon's wavefunction is instantly modified because its left branch component ceases to exist. Photon measurement in the left branch resulted in a *no* photon reading and the measurement *assures* there is no photon in the left branch. As a result of such *no* photon measurement, the right branch photon wavefunction instantly becomes the full un-branched photon wavefunction and subsequent measurement in the right branch will always detect a photon. Although the *no* photon reading instantly modified the photon's wavefunction from branched to whole, the photon remains a virtual photon until actually measured. Therefore, the *no* photon reading is the optical interaction whereas a *yes* photon reading is the reducing interaction that transforms the photon's energy from virtual to real energy. Optical interaction modifies wavefunction shape whereas the wavefunction reduction is always the operation of transformation from the virtual to the real domain.

A photon's full reduction means that light's wavefunction collapses and the photon's energy becomes real in the act of absorption. The photon's reduction is instantaneous and the wavefunction reduces in its entirety across any distance the photon may span prior to the reducing measurement. In a particular interferometer branch the *yes* photon reading happens 50% of the times and in such cases the photon's wavefunction in

Quantum Pythagoreans

both branches instantly and fully reduces even if the ends of the wavefunction are light-years apart. Nonlocal behavior is at the heart of quantum mechanics, and interferometer experiments confirm quantum mechanical behavior at a macro scale. See the Appendix for additional refinement regarding polarization of each photonic branch.

For each device that branches photons, such as the half-silvered mirror and some crystals, the observed light pattern forms as follows: Each and every photon branches and both ends propagate on separate paths toward a screen. Self-superposition may also happen along the way that changes the probability of the photonic presence and the resulting wave pattern then represents the entire photon with its varying probabilities of reduction. At the screen the entire wavefunction reduces in one spot (in one locality) and, if identical photons follow, the wavefunction builds up at the screen with higher light intensity being proportional to the higher probability of photonic reduction. If photons are branched and are directed to propagate toward the screen without self-superposition, both halves are, nonetheless, interconnected in one wavefunction while forming two patterns – one for each branch of the photon. At the screen but one branch reduces the entire wavefunction and the entire photon's energy localizes in one spot. Because each branch has equal probability of reducing, the resulting pattern is composed of patterns or spots of equal intensity.

Slits or gratings allow photons to be branched into more than two branches. Four slits branch each and every photon four ways. You now put a light detector just next to each of the four slits and send one photon at a time toward the slits. You will now make the following successful predictions: One of the four detectors always detects a photon and, since no photons pass through undetected, light's superposition pattern cannot form behind the slits. You can also explain that but a single detector detects a photon because the *yes* photon reading reduces the entire wavefunction at that particular detector. Further still, even though each of the four detectors detects 25% of all photons, you can explain why the photon is always detected by one of the detectors and does not at times slip past all of the detectors: *No* photon readings at the first, second, and the third detector continue to un-branch the wavefunction all the way to 100% at the last

169

detector.⁶³ You can also remove any number of detectors and always make correct predictions. For example, by keeping one light detector at one of the slits and removing the other three, you can predict: **a)** the detector detects 25% of all photons, and **b)** the superposition pattern will be the same as if the slit with the detector was completely blocked because every *no* photon reading at the detector zeroes and redistributes (un-branches) the wavefunction there, and **c)** the superposition pattern will be at 75% intensity because 25% of photons reduce at the detector and do not contribute to the pattern. You can now go back to the light source and remove the requirement of sending in one photon at a time because you know that each and every photon branches independently.

Returning to the virtual electron and to any-thing that finds itself in the virtual domain, there can now be a full closure on the visualization of the wavefunction. An electron is classically perceived as a solid point particle and the introduction of the quantum mechanical representation of the electron as a wavefunction is at times thought to be but an abstract mathematical representation of the point particle's attributes. Classical desire is to keep things localized. Photon wavefunction branching and the detector's optical or reducing interaction with a part of the wavefunction clearly show that the wavefunction occupies different parts of space and that the wavefunction is the actual spatial spread of the entire entity and that the actual spread of the entity can span vast and unlimited distance. The reduction of the virtual entity and the reduction of the entity's wavefunction are one and the same. A free moving electron or other virtual

---

[63] The first detector that interacts with the photon detects 25% of the photons as the detector interacts with one quarter of the photon's wavefunction. In the case of *no* photon reading at the first detector the *no* photon reading un-branches the wavefunction from four to three branches and each of the remaining detectors then has 33% probability of detecting the photon, for any one of the remaining detectors interacts with one third of the wavefunction. The next (second) detector interacts with 75% of photons at 33% probability of detecting a photon and thus registers one third of 75% – that is, 25% of photons. However, if the second detector registers a *no* photon reading then this optical interaction un-branches the wavefunction from three to two branches and the next (third) detector has 50% probability of detecting 50% of photons. Finally, if the third detector registers a *no* photon reading then the un-branched wavefunction is at 100% at the last detector where 25% of photons will always be detected.

particle is spread in accordance with the wavefunction, and the yes-or-no interaction with the wavefunction reduces the virtual particle with the probability that is proportional to the area of the wavefunction subject to yes-or-no interaction. If the measurement engages ¼ of the total wavefunction area then the probability of the entire wavefunction reduction is 25%. Wavefunction can be in up to three spatial dimensions, although some entities such as photons may occupy but two. The probability of reduction is then proportional to the fraction of area or volume of the wavefunction that is subject to the yes-or-no detector interaction. The probability of reduction P_R of a wavefunction is $P_R = \Psi_I/\Psi_T$ where Ψ_I is the intercepted portion of the wavefunction and Ψ_T is the total wavefunction. The reduction is guaranteed when the entire wavefunction is subject to measurement interception. Optical interaction instantly modifies the wavefunction at lightspeed but the wavefunction does not reduce. For light, a mirror always facilitates optical interaction, for mirror does not present light with the yes-or-no construct. The measurement, however, always modifies the wavefunction instantly; it either reduces the entire wavefunction with the probability discussed above or it always reduces but the part of the wavefunction subject to measurement.

More To Lightspeed

Light propagates by relating to entities in space. Light computes its way through space or matter and lightspeed is the speed of light's computations. The speed of the source that introduces light into space is not added to lightspeed: once light enters space it instantly adjusts and propagates at its computing or optical speed.[64] Inside optical materials the

---

[64] The propagation of light in ether is best visualized as swimming in water. The swimmer is launched from a fast boat but as soon the swimmer hits the water he continues to swim at one speed and in the direction of the jump. The speed or the direction of the boat, then, is not relevant to the swimming speed. Light instantly adopts its speed to whatever medium it happens to be in. There is, therefore, no light propagation with ether, against ether, or perpendicular to ether – only in ether. Ether, in the form of uncommitted and spread out electrons, is stationary and does not add to or subtract from lightspeed.

computing complexity increases and lightspeed decreases. Light speed decreases and increases instantly when light enters and exits optical materials the likes of glass. A moving light detector receives light and measures lightspeed to be the same regardless of the speed of the detector. The speed of light is absolute and does not change with respect to any moving source or recipient.

The instantaneous reduction of branched light may span vast distances, but the constancy of light speed and the instant reduction of branched light are not in conflict. Two branches of light originate at one point and both propagate from the point of branching at constant speed. Whatever information or virtual energy light carries continues to be subject to light's computational speed. Un-branching (*no* photon detection) or reducing action in and of itself, and at any wave front of any branch, does not deliver information any faster than the light's actual arrival at the reducing point. Light can reduce only when it arrives there with the information. One interesting aspect is that light branching also parts the information and this effect can be seen as a form of information hedging. Information may be propagating in several directions but there is always only one recipient for any given photon. When light reduces, both the information and the recipient is known even though the information may be traveling in many directions and toward other potential recipients. Similarly, if a single photon is branched and one end propagates in an easterly direction while the other end is stretching in a westerly direction, then such a photon does not carry information at twice the speed of light because it is the same information that propagates at the speed of light from the branching point. To be more accurate about light speed, a better way of saying it is that the speed of information carried by light is independent of the speed of the sender.

In the forthcoming chapters on gravitation the limit on the speed of information will be further refined to a smaller subset of *new and unlike* information where such distinction becomes important for superluminal travel.

Quantum Pythagoreans

Photonic Energy Coming and Going

Light is energy that can do work. When a photon reduces, the resulting energy is real work. Energy, work, and heat parameters all have the same units of measure, and all three parameter names can be used interchangeably. A photon's energy is virtual energy and needs to be transformed to become real energy. A photon transforms at absorption, also called reduction.

Light's path changes in the interaction with orbital electrons in a process called refraction. It can be said that light's interaction with orbital electrons is based on shapes and geometries of both the light and the electron orbitals. If the refracting material such as a prism moves toward light, it will admit light's undulations faster and the refraction will change.

A photon's energy is invariant with respect to any moving or stationary observer and the real work at a photon's absorption is, therefore, absolute. Let's take an object that points and emits light toward the observer. If the observer is moving toward the light source, he or she may theoretically analyze the incoming photons as having higher frequency of undulations and, since frequency relates to photonic energy, the observer may think that light becomes more energetic when the frequency of undulations increases. However, photons of light also span a particular spatial distance and a photon's duration of interaction will change as well. A photon's speed remains constant, and if the observer moves toward it, the photon undulates at a higher frequency but the photon's shorter duration does not make a definitive case whether the amount of energy of the photon has changed. In the illustration below are two representations of an identical photon as it is propagating in the space of a quantum vacuum. The first photon's shape happens when the observer is receding from the light source while the second photon's shape happens when the observer is approaching the light source.

Two photons of identical energy

Since photons cannot and do not impart energy at reflection, the undulating frequency at reflection is not the measure of the photon's energy. The photon's energy does not manifest at reflection or, by definition, during any optical interaction. The photon's up and down undulations are the measures of a photon's presence rather than energy. The photon's shape can change radically when the photon is sent through slits but such change in its undulating shape has no effect on the parameter of the photon's energy because no energy can be added to or subtracted from the photon when it passes through slits. The photon's shape and consequent undulations may be changed dramatically but at reduction the photon delivers but a fixed amount of energy that is not dependent on the photon's shape prior to reduction. A photon can be passed through various and multiple geometries but at reduction the energy will be the same because the geometric forms affect the photon's spatial presence or its path but not the photon's energy.

Moreover, the assertion at the beginning of this chapter also states that the velocity of the observer who reduces the photon is not added to or subtracted from the photon's energy and the photon's energy is absolute. The proof here is logical and based on the conservation of energy. If the photon has a certain amount of energy and the observer has a certain amount of energy, the photon's energy at absorption cannot be dependent on whether the observer is coming or going. If energies were different, then energy conservation would not hold because it takes a certain amount of energy to create the photon and a certain amount of energy to get the observer moving, and then the direction of energy application onto the observer would make a difference at absorption and the energy conservation would not hold. The measure of energy of a photon, then, is in a mechanism that is not related to the shape of the photonic wavefunction.

During the photon's creation the energy of one orbital is subtracted from the energy of another orbital, and the differences between the two energies are transformed into a photon. The energy of the "higher" orbital is greater overall than the energy of the "lower" orbital, but during the photon's creation the difference between the two orbital energy distributions dips below zero at times. Overall, the probability distribution of the photon's energy is positive because the higher orbital has greater

Quantum Pythagoreans

overall energy distribution than the lower orbital. However, as the energy probability distributions are being subtracted from each other at the speed of light, the undulating pattern forms as shown below.

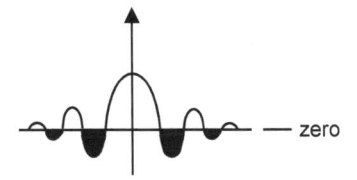

Photon with positive and negative probability

A photonic wavefunction has an inherent zero line. A photon's wavefunction is a probability distribution and the wavefunction carries a mix of positive and negative probabilities. The wavefunction is a probability of the photon's substance, which is a photon's virtual energy. Above the line the energy probability is positive while below the line the energy probability is negative. The overall (net) sum is positive and that is what becomes real energy at a photon's reduction. It is now possible to see that the stretching or shrinking of the wavefunction due to the observer's velocity does not affect a photon's energy because the overall energy of the photon is the net difference between the positive and negative areas. The stretching or compression of the photon does not affect the net difference. At reduction the photon instantly transforms and the net energy is fixed and available regardless of the velocity of the observer that reduced the photon.

Lejeune Dirichlet established the photonic wavefunction mathematically and worked it at the time that preceded the emergence of quantum mechanics. The Dirichlet function is a good example of mathematics being ahead of its application – in this case physics – for at Dirichlet's time the application of his math to light was not apparent. Perhaps even today some may not feel comfortable in equating the Dirichlet function with the photon, but it is likely that the Dirichlet function can be derived from more fundamental physical and geometric constructs.[65] The Dirichlet function also creates plus and minus polarity,

---

[65] Lejeune Dirichlet's photonic function is **$\sin(x)/x$**. When $x = 0$, which corresponds to a location on the vertical axis, this function also contains an example of 0/0. The result of 0/0 is technically indeterminate. In this case,

which may have fundamental consideration in the creation of the positive and negative charge.

A photon can be directed at a pinhole or parallel slits or a crystal and in each of these cases the photon's wavefunction – that is, the photon's shape – will change. The quantity of undulations may change as well and this may seem as if the frequency of a photon has changed. However, the difference in the positive and negative energy probabilities remains the same and at the photon's reduction the overall magnitude of energy will remain the same. A change in photonic shape will change the conditions under which the photon reduces, but the change in undulating quantity or shape is not a determinant of a photon's energy.

Photonic energy is absolute and independent of an observers' movement. At times, however, the observer thinks he or she is detecting photons at frequencies that change with the observer's velocity. In such case the detector is composed of a prism and the approaching light may refract differently because the photonic interaction and consequent refraction may be different. However, while the refraction angle may change, the photon's color does not change and the observer is able to continue the measurement in absolute terms. To a human eye the increase or decrease in speed does not change the color of the incoming photons, but the speed difference between the source and the eye will result in a slight defocus as the refraction through the lens pupil changes.

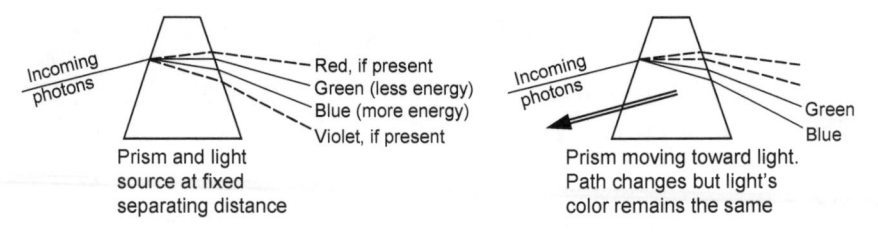

Incoming photons
Red, if present
Green (less energy)
Blue (more energy)
Violet, if present
Prism and light source at fixed separating distance

Incoming photons
Green
Blue
Prism moving toward light. Path changes but light's color remains the same

Refraction increases when prism is moving toward light source

---

however, the value of Dirichlet function at x = 0 is finite and determinate. Guillaume L'Hospital successfully worked the mathematical treatment of a function converging to 0/0 and supplied the test for determinism and the procedure yielding the actual convergent value. In the case of **sin(x)/x** the value at x = 0 is **1**.

176

Quantum Pythagoreans

The change in refraction gives the appearance that photons have different energies but it is only the refracting angle that changes as a result of the speed of the observer or the speed of the source. When approaching the light source, for example, the "light shift" occurs such that the green light refracts closer to the angle of the blue light, while blue light refracts closer to the angle of the violet light. Yet the blue light will remain blue in color. The observer's or the source speed affects the photon refraction but does not affect the photon's energy. Shorter wavelength is proportional to higher energy but only as related to the difference between frequencies that are inherent to two atomic orbitals. Once the photon is created, its energy remains fixed and optical manipulations of a photon such as reflection, refraction, or branching cannot and do not add energy to, or subtract energy from, the photon.[66]

Similarly, a receding light source will produce a lesser refraction but its photons' energy is fixed. It is then more appropriate to talk about undulation change rather than frequency change because undulation of the wavefunction affects refraction but does not affect a photon's energy. The usual analogy of the light wave is with the water wave. Both kinds of waves superpose and form peaks and troughs, but the water wave is a real wave whereas the light wave is a virtual wave. The water wave moves the boat up and down and there is then an obvious linkage to energy since the up and down motion of the water wave indeed produces some work and, therefore, water's wavy motion does have energy. Light waves, however, have virtual energy and light waves change direction through reflection or refraction and without imparting any of its energy on the environment.

---

[66] Some confusion exists among science writers on this issue. It is tempting to say that by approaching the red light at high speed the frequency of the red light will increase to the point of the light appearing, say, green. This is incorrect. The change in photonic wavefunction undulations is not related to the photon's energy and the color of the light will remain the same, even though the color will refract along a different, say green, path. To the observer the red light remains red but becomes out of focus. Light waves and sound waves are different. The sound wave's speed shift relates to the frequency of longitudinal waves but light waves speed shift relates to the change in its refractive path. Additionally, sound waves are real waves because the oscillation is that of real particles (gas molecules).

The Balmer's wavelengths issuing from hydrogen are known to high accuracy and these wavelengths can be observed from within interstellar space. In the laboratory, the refractive shift is zero because the source hydrogen atoms and the refraction detector are not moving with respect to each other. In the laboratory, then, the absolute zero shift can be established. The measure of refraction shift is then useable for the measurement of the absolute speed of your ship: If the interstellar hydrogen from many directions is found to have no refractive shift at two spots along the earth's orbit, then earth's speed will become known in absolute terms. A change in refraction in either direction then directly measures your speed with respect to the stationary ether, because during the calibration the speed of the source is exactly offset by the speed of the detector.

The speed of the light source or that of the detector is not additive to a photon's total energy during generation or at detection – that is, a particular photon will produce a fixed amount of work regardless of the speed of the detector that reduced such photon. A light source will also produce photons with constant work regardless of the speed of the source that produces such photon. In practical terms, a battery in a laser will not drain any faster if the laser is moving fast or slow or forward or backward. Even though different observers may register different changes in the undulation of photons from the same laser, they will all measure identical energy in all photons.

Self-Superposition

A proposition can be made that a standing wave becomes a null entity because the standing wave "subtracts from itself" after reflection and at every point along its axis the standing wave's value is zero. Every individual photon (or electron wave pocket) has zero value at the nodes while at all other points each photon has a unique value. Each value of a photon can be subtracted from or added to with zero as a result if the other wave is in the opposing phase and of equal magnitude. In the illustration below two standing waves are created by reflection but only one adds up to zero. In the case where all photons reflect from a mirror surface, individual photons can and do reduce at positions other than nodes and, upon

Quantum Pythagoreans

reduction, reveal their particular color even though the other photons happen to subtract from them. Even though the photon wavefunctions add or subtract from each other, each and every photon reduces upon conditions that are independent of all *other* photons.

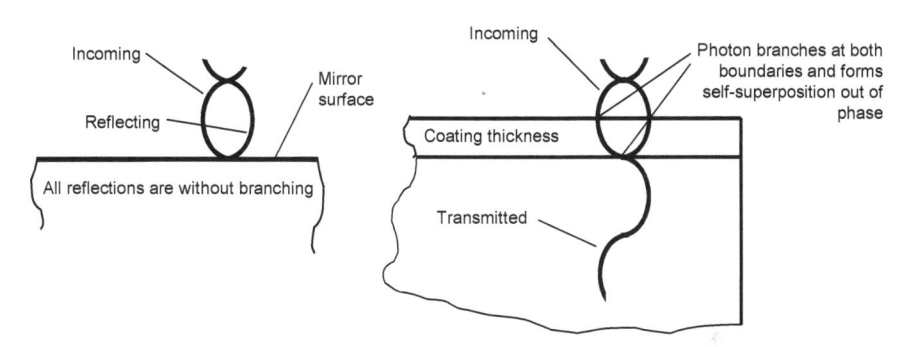

Photon adds to zero if photon branches and forms self-superposition

However, if the net zero of the wavefunction results from *self-superposition*, the photon itself cannot reduce anywhere where its self-superposition is zero. If all photons branch and move such that each and every photon superposes with itself out-of-phase then no photons will reduce at such path. Furthermore, self-superposition may be zero at some path but, since the energy is conserved, other non-zero paths will instantly increase their wavefunction magnitude to compensate for the places and times where self-superposition is zero. This is the case for optical coating where coating thickness is one quarter of the photon's wavelength.[67] In the illustration above, the self-superposition has the look of the *Alice in Wonderland* Cheshire cat whose body disappears but now the cat moves on

---

[67] Optical coating is art as well as science. For example, the index of refraction of the coating material figures in the phase of the reflected arm. The idea is for the reflected component to rotate 90 degrees at the start of the coating boundary while the end of the coating boundary produces another 90-degree reflection that is now 180 degrees out of phase and fully subtracting from the incoming component if the coating thickness is one quarter of the light's wavelength. Moreover, if the angle of incidence is slanted the path through the coating thickness is no longer one-quarter of the wavelength. Different colors have different wavelengths and each photon will encounter the thickness of the coating that is only approximately one quarter of its wavelength.

and away through its wiggling tail. In summary, a photon can reduce if its superposition adds up to zero with another photon but a photon cannot reduce where its own superposition (self-superposition) adds up to zero.

A photon's oscillation usually happens in the geometry of a plane. A photon may branch and the individual photonic ends may go in many different directions but each component of the photon continues to oscillate in a plane. The plane may rotate if the photon optically interacts with matter where plane rotation is no other than polarization. In a quantum vacuum the change in polarization does not happen and the photon can propagate in an infinitely flat plane. If you draw the propagating photon's boundary you will create a strip of a plane. Both edges (boundaries) of the strip are parallel; you have just arrived at Euclid's 5th Proposition of his *Elements*. Should you wish to expand on Euclid you may say the 5th Proposition is valid regardless of the velocity of the observers. A photon that encounters matter and begins to interact by plane rotation is not addressed by Euclid but can be included as follows: For a given straight line (the zero line of Dirichlet) and a point outside the line there exists but one line that is *equidistant* from the straight line. Technically, there are many such lines for different rates of rotation but each rate of rotation is specific to the materia the photon is transiting. For a given materia there is but one line and if a photon encounters no materia then, in quantum vacuum, the 'equidistant' becomes *parallel*. Euclid can also be enhanced by defining a line as a straight path of a point moving from one point to another point. The straightness of the straightedge is then replaced by a photon propagating on an absolutely straight path in a quantum vacuum. The only entities that can move on a curved path and operate in curved and finite space are those entities having mass or charge. A photon has neither mass nor charge. Two side-by-side photons will propagate in parallel fashion on unbounded and straight paths. A photon's path can change only upon interaction with matter.

Euclid defined a point as that which has no parts. A photon indeed has no parts, for a photon can only reduce as a whole. A photon's path is straight because a photon has no charge, and electromagnetic or gravitational fields do not change a photon's straight path. Because the photon can rotate, it may be agreeable to have Euclid's point to be able to rotate either as a dimensionless point or as finite strip of a plane. Novel geometric constructions may be possible.

The Planck Constant

Balmer's equation relates and predicts the plurality of discrete wavelengths of light radiated by hydrogen. In a given experimental setup the particular wavelengths are constantly being emitted and registered by instruments, and it is not apparent whether radiation is continuous or not. However, observations also indicate that different wavelengths radiate at different temperatures and as temperature increases and decreases certain wavelengths start and stop radiating. The presence of particular wavelengths could be measured, but light also has the parameter of *intensity*, which indicates the amount of radiation at a particular frequency. Light intensity also changes with temperature. By 1900, then, it was becoming clear that photonic radiation from atoms is not only in the form of particular and discrete (Balmer's) wavelengths, but also that the source of each and every wavelength is turned on and off. The increase in temperature not only turned on atoms with new frequencies but also allowed more atoms to radiate at existing frequencies. Max Planck introduced a numerical constant as a way of matching up the measured discrete energies of light radiated by atoms and, while its designation is **h**, it is called the Planck constant after its discoverer. Planck also used the word *quantum* in the description of light's emissions. Applying the Planck constant in the equation answered one question while introducing many others. What became apparent is that an electron radiates energy only during transitions between orbitals. Moreover, the non-radiating state at a particular orbital can be explained only by the electron's transformation from a real electron into a virtual electron; the subsequent spreading in radially symmetrical geometry keeps the electron's energy quiescent.

The mathematical alchemist looks into the computable aspects of the Planck constant and realizes that it is not a plain dimensionless number such as π or a dimensionless parameter such as a cycle (as in cycles per second). The Planck constant is a product of energy and time. It is the energy-time product that is constant – the area comprised of energy and time is always the same and so the *action* is also the same. Action may be a

181

nice word but a new label does not explain much. The alchemical interpretation is that it is a product and the Planck constant is then about force. Putting alchemy aside for a moment, from the physics perspective the Planck constant packages different multiples of energy and time that result in the same 'Planck' value. For each Planck value, energy is available in a larger quantity of shorter duration or in a smaller quantity over a longer period of time. The Planck constant does not define the smallest amount of energy but the smallest product of energy and time. Planck formulated the energy carried by light as $E = f \cdot h$ where f is frequency and h a constant.

A family of discrete Planck values is shown in the form of an equation as $k \cdot h/E = 1/f_k$. Incrementing integer k ($k = 1, 2, \ldots$) produces frequencies f_k, which for some values of k are also in agreement with Balmer's equation. The unit of $k \cdot h/E$ is time. Yet it does not all add up because so far there is no explanation on *why* the units of time are quantized and why frequencies are available in increments (multiples) of h that is the Planck constant. Time is, after all, a very accommodating entity and can take on any value and readily becomes zero as well. Also, the atomic environment is symmetrical about a point and all orbitals have associated frequency. The answer becomes apparent when the units of frequency are examined. Frequency is the reciprocal of time – that is, frequency $f = 1/T$ or $1/f = T$ and T is the *period* of the wave in units of time. It is the atomic orbital periods that are quantized.

The electron orbital is explained in a previous chapter as motion that is always radially symmetrical. The electron cannot have radially symmetrical movement if it is concentrated in a point. The orbital electron is indeed the virtual electron. However, energy conservation allows the virtual electron's energy to be visualized in terms of the real domain's parameters such as mass and orbit period where orbit period is the reciprocal of angular frequency. It is the parameter of the *angular frequency* f that lives on in the series $k \cdot h/E = 1/f_k$, and $1/f_k$, having units of time, can have any value as long as it is a multiple of h/E. Any and all references to photons having their energy proportional to frequency refers to the angular frequency of the electron's orbital. The image of a real electron having angular frequency, moreover, is valid for but an instant and only at the instance at the commencement of the electron's transformation – a transformation that happens during orbital change.

Atomic orbitals are separated because electron charges repel, and that forces some spatial separation.[68] But in addition to spatial separation, angular frequencies must each be multiples of specific increments of the Planck constant. Atomic orbitals have two degrees of separation: **(1)** Orbital spatial separation derived from Balmer's equation and **(2)** Angular frequency (orbital period) separation specified by the Planck constant.

A free electron's wavelength needs to match a particular orbital energy if the electron is to be captured by the atom in accord with Balmer's equation. Atomic electrons operate within the structure of the atom and are subject to two degrees of separation. During down-orbital change the electron releases a photon and during up-orbital change the electron absorbs a photon. In either case the photon must have a particular wavelength determined by the difference between the orbitals **$1/f_1 - 1/f_2$** where the orbital **$1/f_1$** corresponds to the higher energy orbital and where but two orbitals can be engaged. Note that the k-th orbital period is proportional to **$1/f_k$** and in accord with the progression of the harmonic series. Not incidentally, this is the same as the notation of ancient Egyptians' fractions.

Ancient Egyptians expressed fractions in terms of the harmonics series **$1/n$**, which always has the numerator of one. The denominator starts at 1, which is in "ground state" correspondence to number **n** in Balmer's equation. Fraction 5/8, for example, would be written as 1/2 + 1/8. Fraction 2/5, however, would not be expressed as 1/5 + 1/5, for in the harmonics series the denominator does not repeat. The expression 1/5 + 1/5 is not allowed, not only because it does not conform to the harmonics series, but also because this expression does not reflect the engagement between two orbitals. The idea is to compose the fraction with particular and unique

---

[68] This is but an introductory explanation. Electrons' wavelengths superpose with each other and then a harmonizing aspect allows but certain ratios of orbital values. The electron "charge" is not an adequate mechanism for separation because the charge is also spread out. Riemann work in spherical geometry deals with this issue and it is likely Riemann work will in time lead to the explanation of the Planck constant.

frequency-based contributors where the first contributor takes the lowest possible denominator (value closest to the "ground state"), and the second contributor takes the highest possible denominator. Fraction 2/5 would then be expressed as 1/3 + 1/15. That is indeed how ancient Egyptians expressed fraction 2/5.[69] Consequently, 2/5 "worth of frequency" is sufficient for two orbitals of 1/3 and 1/15 "worth of frequency." In another example, 3/5 is equivalent to 1/2 + 1/10. All fractions based on binary denominator such as 33/64 are easy to express as harmonics series with term-summing expression of 1/2 + 1/64. Ancient Egyptians treated sub unity fractions as a unique class of numbers and popularized binary fractions with the Eye of Horus mnemonic. Present-day mathematicians reduce all fractions into a real number but for ancient Egyptians fractions belong to a unique class of numbers that are not reducible and are expressed as a sum of two particular members of the harmonics series. In present day notation it would be generally shown as Sum[**1/n**] (for **n = 1** to infinity) and, specifically, 3/5 would be expressed as **0**·1/1 + **1**·1/2 + **0**·1/3 + ... + **0**·1/9 + **1**·1/10.[70] The idea is that mathematics of the ancient Egyptians' fractions conforms to a particular physics phenomenon, which reflects the unique behavior of the atomic orbitals. Similarly, a musician can work with ratios as these relate to the musical octave while being completely oblivious to, and having no utility for, the decimal fractions that the machine operators may need. In fact, the person pursuing the precise cutting and shaping of a stone without hammer or heat does not need decimal fractions.

Following the formulation of his **E = f·h** equation, Planck confirmed that the oscillator that radiates light from inside the atom must start and stop. At Planck's time, some scientists modeled the atom as a continuous oscillator such as a vortex but, because the Planck relation

---

[69] All science writers without exception muse about the "awkward" way ancient Egyptians handled fractions.

[70] Euler worked the harmonics series mathematically and he also obtained expressions converging toward π through particular versions of harmonics series. In the construction of the orbital energy, components need to continue adding up and toward the orbital that includes π.

matched the experimental results, atomic modeling in time yielded to electrons that radiate only during orbital transitions. Photonic energy depends on the difference in two orbitals' energy. The Planck relation drove the new model of the atom despite the realization that the electron staying on in its orbital does not radiate energy – a concept that is impossible to reconcile if you consider the electron to be a point mass electron.

The photon can be seen as a wavefunction that undulates up and down as the photon moves, but such undulation carries the value of the probabilistic existence of the photon and not the photon's energy – even though the undulations are at times mistaken for "frequency." Geometry of the orbital includes 2π, which is encountered every time the photon is created from the orbital. The Planck constant is divided by 2π because the electron's energy that is based on geometry of angular frequency is converted to linear (straight) geometry of the photon at radiation. Number π is the transcendental number that does not contain physical units such as distance or energy. Number π has an infinite mantissa, so the arithmetic way to completely describe π is through a mathematical expression that converges toward π. The number π cannot be drawn exactly as a straight line. π can be expressed as a periphery of a circle of diameter **1** that starts and ends in one point. Saying that a circle having a rational diameter has an irrational peripheral distance starting and ending at one point may seem trivial – yet a potentially infinite number of wavelengths can fit such distance if the wavelength can maintain a symmetry about a point. A circle, then, offers a family of mathematical solutions if such geometry is maintained. Moreover, mathematical solutions exist for the geometry of an ellipse as well.

The Planck constant can also be called the energy-period unity or energy-orbital unit. The Planck constant does not represent an energy-time continuum because, by definition, a continuum cannot be quantized. However, the quantization happens as a result of atomic orbitals rather than through some more fundamental mechanism. In other words, the atom is constructed subject to certain constraints, and quantum energy-period increment **h** becomes a necessity during construction rather that the other way around. In a more mundane analogy, wood and bricks and gym shoes can be technically made in any size but after some thought all of these

●●●○○○○ **The Virtual Domain**

items would become available only in particular size increments (quanta) that make economic sense. It is most likely that other photons can be produced that are not limited by the atomic structure and in such case the photon's energy-period product can become a continuum without the Planck constant. The Planck constant describes the energy-orbital relationship of a photon that is unique to the orbitals of the atom or a molecule. In future chapters on gravitation, the Planck constant is taken as the smallest unit of imparted energy during the creation of the gravitational pull, but, in general, the gravitational unit of energy may be related to another quantum of energy.

Instead of seeing the photonic quantization as some form of a limit or even a barrier, a better way of looking at it is in the form of an opportunity. Photons that do not subscribe to atomic quantization and angular frequency separation cannot be reduced inside the atom and pass right on through.

Any photon is a wave packet and an entity that is indivisible, and thus the conservation of energy applies to the entire photon and without regard to the ways and means the photon is produced. Yet a photon can also be stretched without bound through interacting geometries. A photon can stretch in length but concurrently its average energy density decreases. At absorption, and regardless of the velocity of the absorbing (atomic) structure, the photon will have only the same and fixed amount of energy.

A photon's energy is always virtual unless and until the photon transforms by becoming absorbed. A photon's virtual energy **E** is equal to **f·h**, which is a virtual quantity as well. The Planck constant **h** facilitates the equivalence among virtual quantities of the atom, and the correct Planck expression is then $\mathbf{E} = \mathbf{f \cdot h \cdot} i$. Generically, the Planck constant is the *harmonics constant* of the atomic orbital. Mathematically, any harmonics series needs a particular constant that specifies and creates the next new member of the series. In the case of the electron's atomic orbitals, the Planck constant provides this function.

Superposition of States

Taking light in its full quantum mechanical behavior includes a discussion on the superposition of states. *Yes* photon and *no* photon are two states that are known upon measurement. In each branch of the interferometer there exists 50% of the photon's wavefunction and each half of the wavefunction can be modeled as consisting of a 50% *yes* photon and also of a 50% *no* photon state because, empirically, the photon detector in a particular branch detects a *yes* photon 50% of times while giving a *no* photon reading the rest of the time. Although superposing *yes* and *no* states within each half of the wavefunction explains what is being observed, this is but a statistical, rather than actual, explanation. The probability of successfully reducing the wavefunction is proportional to the area (volume in general) of the wavefunction being measured. In the interferometer, for example, if the measurements in one branch always intercept one half of the total wavefunction then the results will be 50% *yes* photon and 50% *no* photon measurements. The way to visualize this mechanism is that the detector is processing an ever increasing portion of the wavefunction and the probability of the wavefunction reduction is increasing as more and more of the wavefunction is being processed or "consumed" by the detector. The *yes* photon and *no* photons states are not embedded in each portion of the wavefunction. If the wavefunction amplitude is split 25-75 then the branch containing 25% of the wavefunction will yield a *yes* photon measurement in 25% of cases. In the remaining 75% of cases the measurement yields a *no* photon outcome and consequently the wavefunction is no longed branched; an additional detector in the path of the now un-branched wavefunction will always process the entire wavefunction with a guaranteed *yes* photon measurement.

Classical physics deals with definitive answers, which for the most part are obtained by measurement. Searching for a definitive answer, however, ends concurrency because the wavefunction reduces or un-branches. A photon becomes real energy upon a *yes* photon measurement, and a new photon needs to be created before the photon can branch and concurrently move in different directions or superpose with itself again. Compared with the electron, the *yes* electron measurement realizes the same electron that instantly reduces with real momentum while also realizing in one definite position. If the electron remains a free electron then in the absence of physical interactions the electron begins to evolve in space again.

Classical physics is strained in its image and interpretation of quantum mechanics. The realist way of seeing the superposition of states model is that a toss of a coin does not result in a head or a tail but in the superposed state of a 50% head and 50% tail on both sides of the coin. The realist can only muse about the superposition of states by thinking that the twirling coin does not land until the box covering the coin is lifted. It is easy to be confused by this behavior, for if every turn of the real road can only be to the left or to the right, what kind of turn is simultaneously composed of a 50% left turn and 50% right turn? Well, light can get to a fork in the road and go in both directions at once, thus hedging the decision to turn until one of the light's branches has a *yes* photon interaction or until one of the light's branches has a *no* photon interaction with a detector. Light can also go 50% to the right and 25% in two additional branches on the left at the same time. A parted branch of light can meet its own previously separated branch where such a meeting becomes self-superposition. By now the obvious distinction is that only the virtual entity can have various parts of its wavefunction in different places at the same time and such parts can even superpose on top of each other with up to three degrees of freedom. A photon is a virtual entity and it is quite natural and easy for light's branches to go in separate directions all the while remaining an altogether whole entity.

Inside the interferometer the branching of a photonic wavefunction results in two well-defined crests but this is but a specific case that is facilitated with a particular kind of instrument. An electron's transition into the virtual domain, on the other hand, results in the electron's position becoming spread out over a contiguous volume of space that can have highly nonlinear or pierced shapes. A single, zero-dimensional point in space – the real electron's position – becomes the volume of the wavefunction. The wavefunction is the entire particle in virtual form and to guarantee a yes result upon measurement the measuring instrument must interact with the entire entity such that the entire wavefunction passes through the detector. In the case of a photon, to guarantee detection of a photon in the interferometer requires two detectors – one in each branch – to fully intercept light's wavefunction. Even in the worst case the detector pair intercepts the entire wavefunction and the photon will always be detected. Alternately, both photon halves can be reflected with mirrors onto one path; a single detector will then always detect a photon because the detector processes the entire wavefunction.

The virtual domain and the use of virtual numbers define each other. All i-based numbers are virtual numbers. The virtual domain is also unique in that the wavefunction branching and superposition forms there. Any and all virtual variables or virtual numbers have the property of instantaneous reduction. A virtual diffusion coefficient of the Schrödinger equation directly leads to the evolution of the wavefunction in an adjacent volume of space while having the ability to reduce instantly.[71] A diffusion equation with real diffusion coefficients, however, describes the actual movement of individual particles into space such as is the case with perfume, or the penetration of ion particles into another medium such as silicon, but such particles remain real and localized with a particular set of coordinates.

It is now also apparent that the real and virtual domains form a duality where the real and virtual must be separated *and* bridged in some way, and, quite importantly, that the real and virtual domains cannot be merged into one domain because they each have a different computational modality. In the upcoming chapters the duality is explored in the framework of organizational growth. The two halves of our brain will each do their respective best for the purpose of an overall understanding of the duality.

In summary, light is a virtual entity and light's wavefunction is the actual distribution of light's virtual energy. When the photon of light is absorbed, light transitions to the real domain at one locality and light's virtual energy becomes real energy at the midpoint between two real entities that receive the photon's energy. The real and virtual energy are two forms of energy and energy is conserved in its real *or* in its virtual form because the transitions between the real form and virtual form happen in the framework of energy conservation. Applying mathematical group

---

[71] Generalized equation describing light diffraction contains virtual components associated with glass absorption and this also shows that photon reduction is instantaneous and a portion of photons dissipates as heat inside the glass.

theory to the photon, a photon's energy transforms from virtual to real energy upon the operation of absorption while the optical operations of reflection, refraction, branching, or a null reading at detector (unbranching) keeps the photon's virtual energy invariant as virtual energy. Optical interaction is thus defined as a modification of the photon's wavefunction where the photon's energy remains indivisible virtual energy. The operation of absorption always results in real energy such as movement or pressure or electrical energy that ends a photon's existence while the conservation of energy holds. Finally, the answer to 'Does the wave of a photon behave as particle?' is answered by: 'Yes, with two bodies.' When the wave of light touches two bodies it gets them moving in an outbound fashion and both bodies expand. A photon of light transforms – that is, the photon is absorbed – *always and only* when the conservation of momentum is upheld; a photon does not transform under any other conditions. A photon can only transform as a whole. If the framework of momentum conservation exists then momentum has priority and real energy manifests. Movement has priority over light propagation if the framework of momentum conservation is offered. Movement then always results between two bodies but such movement is not analogous to a collision of one particle with another and cannot be compared with one billiard ball colliding with another. If matter is confined then a photon's absorption manifests as increased pressure where the absorption of light converts light's virtual energy into real pressure in the analogy of a spring's compression.

When a particular attribute consists of but two qualitative states, such as energy having a real or virtual state, the group theory calls the dual states *parity*. Energy is also conserved through parity and this means that a particular quantum of energy can be fully real or fully virtual, while the act of transition between the two always conserves energy overall. There are many different dualities and parities that come naturally, such as the head or tail of a coin, that are two states of the same coin. Every number also has an even or odd parity where the parity may be conserved or transformed through numerical manipulation. Through experimentation it is now natural, though not obvious, to think of the independent and stable atomic particles as having a positive or a negative charge. Parity, then, is the fundamental parametric symmetry of any entity. It can be seen that if the transformation is a law of nature then the parity of an attribute guarantees transformations to happen.

The existence of virtual energy may be difficult for some to accept. Once the opportunities of virtual energy begin to appear, however, the concept of real and the virtual energy parity will become second nature as well.

Light, Just Warming Up

In a laser, external energy is transferred to electrons that rise and occupy a known level – in a process appropriately called pumping. When electrons spontaneously drop to a lower level, photons of a particular and nearly identical frequency are released. These photons are formed by a unique process under particular geometries and propagate out at lightspeed. These distinctive photons can relate and interact with processes that are similar to those that generated them – other atoms. Detection of such photons requires materials that absorb them, and our eyes also do an excellent job for absorbing and seeing many photons of many frequencies.

As complex or mysterious as atomic geometry may be, the atomic formations that create, absorb or optically interact with photons are but a small subset of all possible photonic formations. There is no property of completeness that is associated with atomic photons. Any accelerating charge will radiate energy in the form of a photon. Just as there is no limit on the generation of photons, there is no limit on the construction of structures that would intercept – that is reduce – such free-form photons. There is no limit on geometric configurations that would interact with some photons but not with other photons. There is no reason why photons cannot be wavefunctions of complexity and shape that, for example, would and could be absorbed only by a unique structure. There is no reason that a photon shape could not describe an entire and useful word such as 'stop.' By extension, a single photon can be formed to represent an entire sentence – although a unique structure would then need to be built for each and every sentence. It is possible to surmise that some photons pass through metals because the particular photon's geometry and the metal's atomic geometry do not interact in a way for the photon to reduce or reflect. There is no reason to be stuck in our perceptions just because photons we call "light" or "radio waves" reflect from metals and, therefore, all photons must reflect from metals. Perhaps what some think to be ironclad laws of physics may turn out to be presumptuous, unwarranted, or narrow-minded extensions of laws of physics that operate in but a small subset of the real domain. Furthermore, something may appear "metallic" but it will not necessarily reflect photons from radar, for photons emitted by radar are very particular photons that reflect from metals listed in the periodic table of elements as metals. A photon, being created by an accelerating charge, can be visualized by placing an electron at the tip of the wand. You now have three degrees of spatial distance to move the wand, while the acceleration and/or the curvature will determine the energy of the photon. The length of your wand waving also determines the start and the stop of the photon. Your custom made photon, then, exists with three spatial parameters having a unique energy profile between the beginning and the end.

Lightspeed is the native photonic ability to pass through uncommitted electrons of ether, which provides the simplest possible interaction. Any interaction of light and other matter, even if light passes right on through, will slow light's propagation on account of relational computation or comparison between light and matter. Presently, light's speed slows down significantly in some materials at very low temperatures. More importantly, though, new opportunities exist because light interacts with matter even though matter may be in its virtual state. Light interacts with virtual matter even though virtual matter is invisible. Further, virtual matter is not pulsing with gravitational eigenstates and virtual matter is not subject to, nor does it produce, gravitational force. In any case, light is not influenced by gravitation. Both light's wavefunction and gravitational wavefunctions superpose in space without interacting. Light's interaction with matter is dominant, and particular photonic shapes will bind virtual matter such that matter may become visible to some extent while possibly remaining free of gravitational forces. Light refracts when optical interactions happen with virtual matter because the interacting light's speed decreases and its direction changes. Light can also reflect when optical interactions happen with partially real matter.

Mechanical compression of gas releases longer wavelength (heat) photons but most photons get reabsorbed within the gas and that is noted as an increase in gas temperature and additional pressure. Some portion of heat photons continues to leave the gas volume while the rest interact in the system. Continuous absorption and radiation of photons can be seen as recycling, rework, or regeneration of photons, which manifests as heat conduction. When a gas molecule absorbs photons, electrons' orbitals are rearranged and orbitals can also include more than one atom. Electrons that form orbitals spanning more than one atom are designated the valence electrons. Heat photons add energy into a web-like valence electron arrangement that keeps atoms interconnected in a molecule. It is also apparent that the absorption of every heat photon increases the separation of the atomic pair, which results in a greater gas volume. If gas volume is being held constant, the absorption of heat photons increases the pressure within such volume.

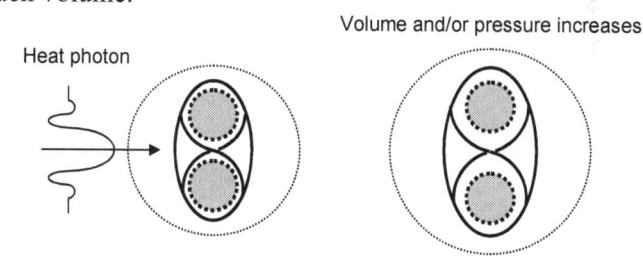

Absorbed photon adds real energy to valence electrons

The incoming photon with a particular shape finds the gas molecule in a molecular structure that facilitates photonic reduction and, therefore, photonic energy is included in the molecule. A similar or "likeable" molecular structure is a structure that reduces photons with similar geometries, much like the electron that is captured inside the atomic structure because the electron finds a computable environment there. A molecule that is absorbing heat photons can be likened to a programmable spring. The spring can be strengthened when photons are absorbed into it and gas volume and pressure rises. If the pressure is too high, however, excess photons (heat) leave the molecule, the spring weakens and pressure decreases. There is no reason why novel photonic shapes could not entice electrons into forming new and possibly tighter electron orbitals. New molecular formations are possible through light sculpting. New atomic formations are more likely to happen if the atom remains in the virtual domain free of gravitational forces. Yet, core sculpting with computational photonic shape methods may also be possible.

The opportunities are unlimited. Photonic shapes could be created that pass through water almost unhindered. Customized photons can be sent for detection by particular systems while other systems would not intercept such photons. Presently, a message containing some knowledge is chopped up into bits and transmitted as a series of ones and zeroes over a selected physical channel whose sole purpose is minimal attenuation. The message is then reassembled to present the recipient with the content of the message. The question for years to come concerns the possible existence of innate ways of receiving and expressing knowledge as various shapes of photons that are constantly present all around us. While detractors may refer to the virtual domain simply as chaos, the question for the future concerns the existence of large amounts of virtual energies that exist in organized form and that are readily transformable with particular frequencies into real energy through geometric means.

Synaptic connections between brain cells interconnect axons and dendrites, and are generally perceived as on-off switches. In an apparent analogy to computers, the switch is 'on' to allow the passage of signals, or the switch is 'off' and no messages flow. Because of the computer model, the research effort centers on how chemicals, ions, photons, or electrons get across the synaptic gap and turn the synaptic switch on and off. Quantum mechanical approaches are also used to analyze the width of the gap to assess how the electron could possibly tunnel across and affect switching. The analysis of a gap, however, shows the gap to be too wide for the quantum mechanical effects to take place. This may be true but the general conclusion that quantum mechanical effects play no role in synaptic interactions are based on the narrow model of the synapse as a switch that is turned on or off via the inside channels between axons and dendrites. This is perhaps the most splendid example on how "scientists" think and how they cannot get outside of the box – in this case the mindset of the computer box. Synaptic connection is a gap having a unique profile that exists to reduce particular external photons – that is, synapses intercept information that is available in photonic form and that reaches the synapse through air, bone, and brain matter. Similarly, the learning process is a process of creation of a unique and particular synaptic profile that is then ready to intercept the learned knowledge. A single photon or a photonic conglomerate that represents a sentence or a "rounded" or a coagulated piece of knowledge is intercepted and reduced as a whole inside the synapse. In the human brain many synapses are created in the process of learning but many also wither away if, for one reason or another, synaptic connections are not exercised and photonic energy does not reduce there.

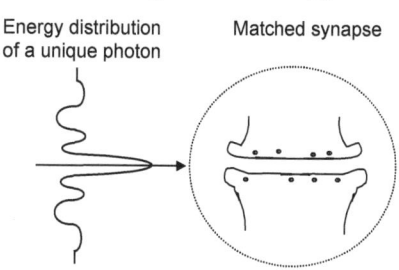

Photon absorbed in synapse programmed for particular photonic shape

The synaptic profile is uniquely created during learning, studying or practicing, and particular factual information is stored in the shape of the synaptic profile, possessing innate information rules. Associated relevant information and actions are generated when photonic information reduces at the programmed synapses thus interconnecting and activating various cellular networks. Also interesting are the synapses of the amygdala that have high priority and intercept "critical" or "action" messages. Because the amygdala cells have power to freeze or incapacitate muscles and produce stupor, amygdala may be targeted if instruments are directed at the lower back side of the brain. Neural cells of the hippocampus have a unique triangular shape, which leads to a proposition that the function of the neural cell is to facilitate particular formation (programming, profile building, gap shaping) of synapses. In general, there is no single or simple way of controlling the human thinking process *en masse*, because different data work with different people. At the same time, there is no easy way to isolate the brain from innate information through a physical barrier because the photons will not reduce within geometric structures of everyday matter.

Overall, the synapse is a content-addressable switch. Content addressability is the grand prize of computing, as it dispenses with *ad hoc* searches that are inherently sequential and require multiple passes across the entire database. Moreover, ongoing sequential processing is likely to encounter intractability.[72] Musings about the human brain include the slow speed of information propagation through the axon – particularly when compared to the affordable desktop computer where data streams are a million times faster. Content addressability goes a long way toward having the slow brain come on top of a fast machine. Photons, then, can be and are created for content addressability. Photons are ideas and knowledge, and such knowledge is accepted and understood by structures, be they organic or inorganic, that have some affinity for it or that derive some benefit from it. The match need not be exact, for the photonic reduction has some positive probability of actually happening and such probability is

---

[72] Left brain processing emulates and manages causal systems. At times, encounters with intractability forces the brain to switch to right brain centered processing. This change can be quite dramatic because it requires complete reorganization of the thinking baseline from *if-then* to *what else* modality. Real and repeatable sensory input to the brain is supplemented with a vast and possibly infinite quantity of weak signals that must be processed circumstantially. In about one-third of the cases the transformation is successful and the person stabilizes without drugs on the relationship-based right side of the brain. In the second one-third of cases the person stabilizes by retaining the left brain focus while able to transition, usually willingly, to right brain centering when necessary. In the last third of cases the person is unable to stabilize. Reality-driven processing happens in the left brain and the sensory input is primarily from the usual six senses of hearing, sight, tactility, temperature, taste and smell. (The ability to detect and classify temperature is an oft ignored but real human sense.) However, content addressability of the synapse also facilitates the receipt of information from photons that enter the brain circuits from without and in addition to the familiar sense organs. A person is usually not cognizant of the fact that the information is reaching the brain directly from the virtual domain, and the information is treated as if it is coming from sources that are real. The large quantity of incoming data's relationships and associations is highly divergent and the lack of stability is at times likened to 'engaging the dragon,' 'dying,' 'visiting the dead,' or as 'schizophrenia.' If the person stabilizes, the outcome is also described as being 'second born,' 'slaying the dragon,' 'becoming a shaman,' or as 'having no recurring episodes.' Stabilization can take years because the conversion happens in stages with a corresponding sense of each time having your brain turned "inside out" and without any hint of finishing. While it may be impossible to resist the onslaught of information, it is also up to the person to figure out when enough is enough because intrinsically the relationship formation is infinite and does not stop by itself. Within the shamanic tradition the lack of reality and subsequent "illness" is anticipated and the stabilization statistics are much more favorable.

increasing with increasing likeness. The down side is that overly specialized synaptic constructions will not receive knowledge in other areas and may in fact wither away if no photons are forthcoming.

There are both the chemical and electrical triggering mechanisms associated with synapses. There is no reason why structures could not exist that would intercept free electrons with particular shapes and vivibrations. In the human brain such electron reducing structure could possibly result in a buildup of charge. However, it would be a very clean model to have the reducing virtual electrons trigger the electrical synapses while, also, the reducing photons trigger the chemical synapses.

"Dad..."

Yes, Alex.

"When one body moves it gains momentum. It also sends a signal to the other body but that takes a while and the bodies are not synced up to conserve momentum at the same time. So, if bodies are far apart, how can you possibly conserve momentum and give momentum to both bodies simultaneously?"

That sure is the classical way of looking at it. Quantum mechanically, gravitational wavefunctions exist in superposition, one together with all the others, and they interconnect both bodies. When some wavefunctions instantly reduce, each body receives an equal and opposite amount of momentum.

"But that means that momentum appears out of, ... nothing. That means that there is some knowledge out there that figures out how much momentum should appear – and when and where it should appear!"

So?

"Hold your quantum leaping lizards, Pops. Nothing can go faster than the speed of light."

Gravitational wavefunctions are between bodies. When wavefunctions relate – that is, when they compute as a result of changing distance – there exists a computational result for new momentum realization that is constantly changing. Yet it is only a potential. When the wavefunction does reduce, it happens instantly and the imparting of momentum to both bodies is instantaneous and in agreement with the conservation of momentum going all the way back to Galileo and Newton. The information is in place already – at both places already – and the wavefunction reduction between both bodies has no delay.

"If I think about a planet and how it can change its path going around in an elliptical orbit and how it can speed up and slow down and speed up again, then there must be some momentum increase and decrease."

That's it! The orbit can be described with an equation but you are thinking about the mechanics that bring it about.

GRAVITATIONAL WAVEFUNCTION	200
Linkages	*205*
Two Components of Matter	*210*
Antimatter	*211*
THE SOURCE OF TRACTABILITY	215
GETTING THE MOVE ON	218
Spatial Distance Takes All	*219*
Force Muscles In	*220*
New Definition of Inertia	*222*
Force Takes Another One	*227*
Mass Weighs In	*230*
2D Is Flat	*233*
3D Is Sun	*234*
MAPPING HYPERSTATES	235
Force and Mass Take It In A Ratio	*237*
Ring and Rings	*239*
Summary	*241*
ALCHEMY	243
Circle	*253*
Intellectual Property	*255*
CORE	258
Continuum of Energy and Direction	*259*
Projection	*260*

Gravitation

Kepler proposed that the moon is responsible for ocean tides, for he thought that a solid massive body and a body of water are qualitatively the same for the purpose of gravitational attraction. Galileo did not like Kepler's moon tide theory, partially because he disliked Kepler's alchemical 'moon is moist and attracts water' mindset, which Galileo considered, well, unscientific. Galileo did not play the role of a debunker, however, for he had a theory of his own. Yet, Galileo's notion that earth's rotation causes tides did not pan out and Kepler's theory of gravitational force-at-distance continued as a viable proposition.

●●●●○○○ Gravitation

Gravitational Wavefunction

"I got a brand new pair of roller skates,
you got a brand new key"

Melanie
Brand New Key

In its simplest form the gravitational link includes but the linear momentum at each end. At minimum two atoms are needed. It is likely that neutrons compute and realize the spin and, if so, basic hydrogen atoms can accumulate but cannot commence spinning. In general, the gravitational link consists of both the linear and the angular components. To conserve momentum, both ends of the gravitational wavefunction consist of equal amounts of virtual energy that becomes the moving energy upon reduction while the linear momentum is always realized in the inward or contracting or centering direction. Similarly, the angular momentum sums up to zero, which necessitates one body to spin clockwise while the second body spins counterclockwise. The reduction of the gravitational wavefunction is periodic. Force periodically appears as the gravitational wavefunction reduces. If the resulting force is in a direction that is also a degree of freedom, the body commences moving in that direction of freedom. Unlike the wavefunction of a free electron or light, the virtual energy contained within the gravitational wavefunction localizes as momentum and realizes equally at both ends of the wavefunction.

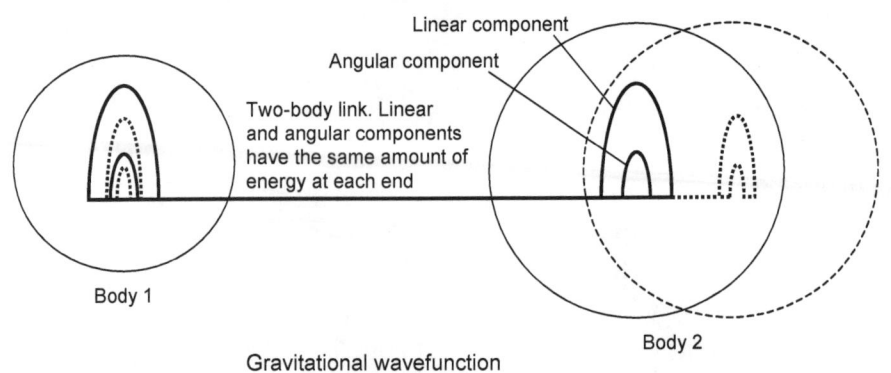

Gravitational wavefunction

Body 1

Body 2

Linear component

Angular component

Two-body link. Linear and angular components have the same amount of energy at each end

Gravitational granularity is that of the proton and neutron. While the gravitational attraction and mechanical force both produce acceleration, mechanical force does not and cannot act on each and every nucleus in an atom of the body it is accelerating. An inflated balloon, for example, will become deformed under mechanical force that accelerates it but gravitational force will accelerate the balloon without deformation. Mechanical force and the force of gravitation are thus qualitatively different forces because mechanical force is not partitioned and applied to each and every atom. Mechanical and gravitational force may produce identical acceleration but the speeding up mechanisms are qualitatively different. A human body subject to gravitational force will feel no stress or pressure on the body as the body is accelerating. Consequently, a human body can withstand substantially greater acceleration from the force of gravitation than from a mechanical force.

In previous chapters the discussion of a wavefunction covered both the electron and the photon. Gravitational wavefunction is a bit different. A free and stationary electron wavefunction spreads continuously and may be bounded by light. A free electron's wavefunction reduces at the attempt of parting the virtual electron's volume into a plurality of discrete volumes of space. A free electron's wavefunction also reduces when interacting with light or a detector.

A light's wavefunction propagates at its computational speed that is a function of the medium such as ether (quantum vacuum), air, or glass. A photon wavefunction reduces at any time an outbound momentum can be created in the framework of momentum conservation.

A gravitational wavefunction links physical bodies while the magnitude of the virtual energy changes at lightspeed with the distance between the bodies; it reduces periodically while the amount of the imparted energy is a function of the separating distance between the bodies.

If separation between bodies increases, the amount of the virtual energy that is available for reduction decreases equally at each end. At some point a question arises as to the minimum amount of the available virtual energy that resides at each end. It is not difficult to propose that the amount of virtual energy can be arbitrarily small, particularly since virtual entities can exist in infinite superposition and also because the photon's wavefunction can be arbitrarily small when the photon parts via the half-silvered mirror. Yet when it comes to the actual realization of momentum at the atomic level, it becomes apparent that quantum states in the atom cannot accommodate an arbitrarily small magnitude and arbitrarily fine resolution of energy. The amount of virtual energy may be continuously recomputed as distance between bodies varies, but while the virtual energy can add or subtract various wavelengths that comprise energy, it is tenuous to propose that the computational resolution at the quantum level of the atom has arbitrary precision.

The energy separation between Balmer's principal orbitals greatly exceeds the Planck constant. Transitions of electrons between other quantum states produce discrete quanta of radiating (light's) energy, and in the second degree of computable separation the atomic structure produces the absolute minimum of energy quanta. Yet these quanta are inherent to the geometry of the electrons' orbitals and not that of the core. Presently, there is no theoretical framework for assessing the smallest quanta of gravitationally realizable energy. Presently, the Planck constant defines the smallest amount of energy only because it matches up with experimental results associated with electrons' orbitals. Gravitational wavefunction compels the question as to the minimum quantum of gravitationally induced momentum. Should this amount be identical to the Planck constant, it gives the Planck constant an additional purpose. On a simpler and practical level, the Planck constant is related to the precision of the computer's register because the register capacity, which is also its precision, defines the smallest amount the computer can *differentiate from zero*. The physical interpretation of the Planck constant is that an energy value for any real thing starts at *near* zero and this differential is equivalent to the value of the Planck constant. The Planck constant defines the smallest amount of any real thing. Should a computer be built that differentiates to the level of the Planck constant then it would be close to the precision of Mother Nature. This computer would be able to manage a trillion dollar amount down to one trillionth of a penny in one arithmetic operation. Mathematically, gravitational function is the *even* function and the intrinsic even function characteristics such as infinite superposition, the 50-50 energy distribution about the vertical axis, and the absence of mass are indeed among the parameters of the gravitational wavefunction. Gravitational wavefunction is a virtual entity and may well be called the *gravitwin* or the twin peak attractor.

There are now two significant characteristics of the universe that can be answered. Because the gravitational wavefunction as well as the light's wavefunction are virtual entities, photons of light may superpose with gravitwins but they do not influence each other. Second, when the separation between bodies becomes so great that the gravitational wavefunction does not reduce on account of not having enough energy, the gravitational pull ceases and the protons and neutrons of the core of matter do not reduce – that is, matter remains in its virtual state because energies cannot reduce in fractions of the smallest gravitational quantum. The material universe, therefore, can grow and expand in its tangible form only from within. The question, "What is at the end of the universe?" then offers the answer that mass must exist at some minimum density to sustain itself, while the organizing and non-chaotic aspect of matter is an additional component of matter sustenance yet to be discussed. The far end of the universe is growing at the rate at which matter can organize and sustain itself in its real form. It should be a good guess that supernovas happen with greater frequency at the edges of the forming universe because the organizing knowledge is at its weakest there.

A branched photonic wavefunction is unbounded in distance yet the gravitational wavefunction is bounded to a distance maximum that may be linked to the value of the Planck constant. The qualitative difference is that the energy of a photon is of fixed and known and constant amount whereas the gravitational energy available for momentum realization is a variable. Gravitational energy must be continuously recomputed. A photon wavefunction, on the other hand, can spread without bound because the particular photon's energy is fixed and is, therefore, not a function of distance. In the Pythagorean tradition the universe runs on numbers or, more appropriately, on tractable computing that applies numbers' unique properties. The update on the Pythagorean view is that the universe can construct any structure if the structure can remain tractable with the mechanism of transformations between the real and virtual domains. Computations can include any and all parameters such as energy, spatial distance, spatial direction, symmetry, and geometry. For example, the energy of a reduced photon can be computed at any time and at any speed because the energy of each photon is fixed, and the distance between light's branches is not relevant to the computation because the light's wavefunction reduction is instantaneous. Gravitational energy available at both ends varies, and can be as small as the value of some constant, possibly the Planck's. Gravitational energy becomes known at a particular value at the instance of the gravitational link's reduction.

Although a free electron's charge can spread, an electron's charge cannot be partitioned and thus lowered, for any attempt at physically cutting the wavefunction of an electron reduces the electron into a point. The speed of light is also absolute and finds its maximum in the ether. The three parameters of Planck constant, electron charge, and lightspeed can be arithmetically combined to produce a quantity that has no associated physical units such as meters or seconds – that is, the result is a plain number that is very close to 137, an incomposite number.[73]

Overall, photonic wavefunction reduction results in real energy but such reduction can happen only during interactions with matter. Electron wavefunction reduction results in real mass of the electron and the reduction can happen during interaction with light or matter. The reduction of the gravitational wavefunction results in inward movement while the reduction is periodic.

Gravitational wavefunction periodically reduces and accounts for linear and angular movement. A satellite in orbit is subject to net gravitational force that keeps the body at about the same distance from the center of the body it orbits. A body in orbit is weightless but is nonetheless under gravitational force that is nearly the same in magnitude to that at the surface. A body in earth's orbit is not in "microgravity" but is subject to a gravitational force that continuously accelerates the body toward the Earth. The orbiting body has a tangential velocity large enough not to collide with Earth. The point, however, is that despite weightlessness the body is subject to the gravitational force at all times. If a body is at the center of the earth it is no longer being accelerated by earth's mass, even though it is still subject to gravitation from other bodies such as the moon. Similarly, at a point close to the moon where the earth and moon's gravity neutralize, there is then a true absence of the gravitational force due to both the earth and the moon but not due to the sun. The true absence of the gravitational force means that the atomic core is not being materialized because the gravitational force is not reducing there, and such conditions are conducive

---

[73] Number 137 derives from the expression $\mathbf{h \cdot c / (2\pi \cdot e^2)}$.

Quantum Pythagoreans

for matter to remain in its virtual state. A mass body growing in size then also has the capacity to form a hollow core, even though the core would continue to have matter in its virtual state.

Linkages

When two entities relate and influence each other, a link necessarily exists between them. A link that is not directly visible is a virtual link. Force of gravitation manifests by attraction and by torque, but the actual link between two physical bodies is not directly observable, leading to the premise is that the gravitational influence or the gravitational link is a virtual link that exists between atoms. A virtual electron wavefunction and light wavefunction can reduce instantly and so can the gravitational link. Gravitational links interconnect bodies and link reduction imparts an equal and opposite linear momentum as well as an equal and opposite angular momentum to each body at both ends of the now reduced link. Having defined the link as invisible and therefore virtual, the link's energy is virtual energy as well. The reduction of the gravitational link transforms virtual energy into real moving energy.

The instant nature of the gravitational link reduction resolves the long-standing mystery of gravitational speed. Present consensus calls for the speed of gravitation to be much greater than the speed of light. Indeed, taking lightspeed as the speed of adjusting interplanetary orbits would result in oscillations because for each change there would be a delay in force compensation. Such oscillations, however, are not observed. The orbital equation is a mathematical result that does not say much about the actual underlying mechanics of motion because the equation is a computable result of interactions among most, if not all, relevant variables.

The instantaneous link reduction signifies that a particular event, action or a manifestation is not a function of distance. When distance is not needed to describe momentum creation, another serious challenge to classical reasoning results because the variable representing spatial distance disappears. The amount of momentum creation does depend on distance but the operational principle of gravitation is nonlocal and independent of distance – provided the available energy is in excess of some constant.

205

● ● ● ● ○ ○ ○   Gravitation

The magnitude of links decreases with the square of the distance between bodies and this allows a drawing of a boundary of relevance. The boundary is relevant and sufficient if bodies' mathematical description inside the boundary matches the actual bodies' movements. The boundary of relevance then allows all bodies outside the boundary to be ignored with the apparent benefit that multi-body gravitational assessment does not need to include all bodies in the universe.

Up to this point the visualization of gravitational links is simplified by dealing with two bodies at a time. Gravitational virtual links among three or more bodies are not multiple one-to-one links because individual atoms do not partition or time-share gravitational interaction between one body and another body. However, because the gravitational force has linear properties the overall gravitation is indeed a sum of all individual gravitwins among all atoms. Many-atom gravitation is computed from among every pair of atoms and results added up to obtain the overall gravitational wavefunction: both the magnitude and direction. Links between gravitationally interacting bodies form and the link at any one atom is a superposition of all possible links. In a three-body system, for example, any particular body has a net gravitational link as if the other two bodies merged into a single body[74].

It is apparent that Kepler thought of gravitation as directly proportional to mass and inversely proportional to distance. His conclusion on gravitation is qualitatively correct if quantitatively incomplete. In the throng of the Counter-Reformation, Kepler was the first person to successfully apply the art of the mathematical formula to orbits – and to this day Kepler's equations are used for calculations of satellites in orbit around the earth. Kepler created a new discipline of using a mathematical equation that describes something as strange as action-at-distance. In his analysis of ocean tides, Kepler realized the tides are mostly the result of the moon's gravitation and only marginally affected by the sun's gravitation. It was apparent to Kepler that not only was the gravitation between sun and earth stronger than gravitation between sun and moon, but the timing of the

---

[74] If the second and third bodies have chaotic movement the merged body continues to be chaotic. Similarly, if the two bodies have computable movement the merged body continues to be computable.

tides also indicated that the moon exerted the greater gravitational force on the earth.

Newton's improvements were two-fold. First, he postulated and later verified that the force of gravity decreased with the square of the distance. Second, and possibly more importantly, Newton applied a summing approach to gravitation that allows each pair of bodies to have their own mutual component of gravitational force while the overall force becomes the sum of all individual components. The summing of forces mechanism is called linearity while the discipline of summing forces in space became known as vector arithmetic. Newton's two improvements earned the *universal* label for his law of gravitation. The summing mechanism is no other than the superposition of gravitational wavefunctions.

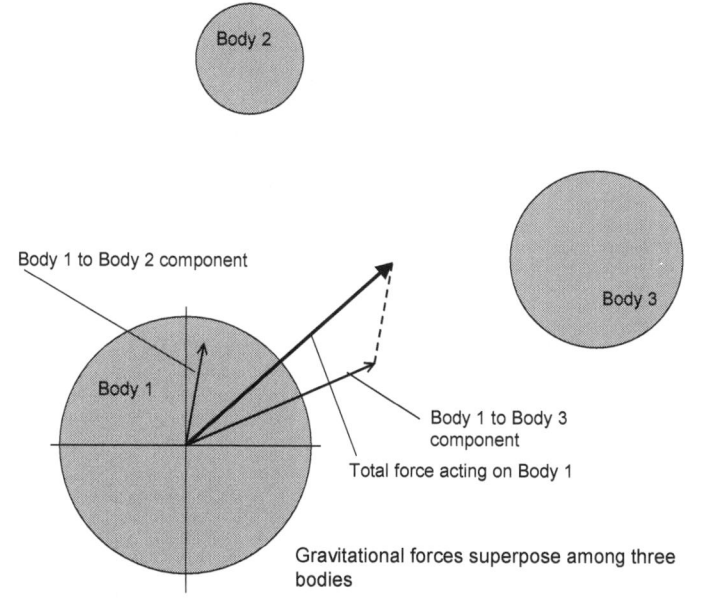

Gravitational forces superpose among three bodies

In a two-body system, a force at either body is proportional to the product of the masses of the bodies and indirectly proportional to the square of the distance between the two bodies. Forces are equal in magnitude and opposite in direction, and as vectors they sum up to zero. In a three-body system the net forces are generally not equal in magnitude, but as vectors they continue to sum up to zero because each and every mutual *component* of the vectors is equal and opposite in direction. Consequently, and if the conservation of momentum is to hold, the

●●●●○○○ **Gravitation**

reduction of the gravitational wavefunction among three or more bodies also produces momentum that sums up to zero. In the illustration below it is easy to verify that the components of force between body 1 and 3 add up to zero, body 1 and 2 components also add up to zero, and body 2 and 3 components add up to zero as well.

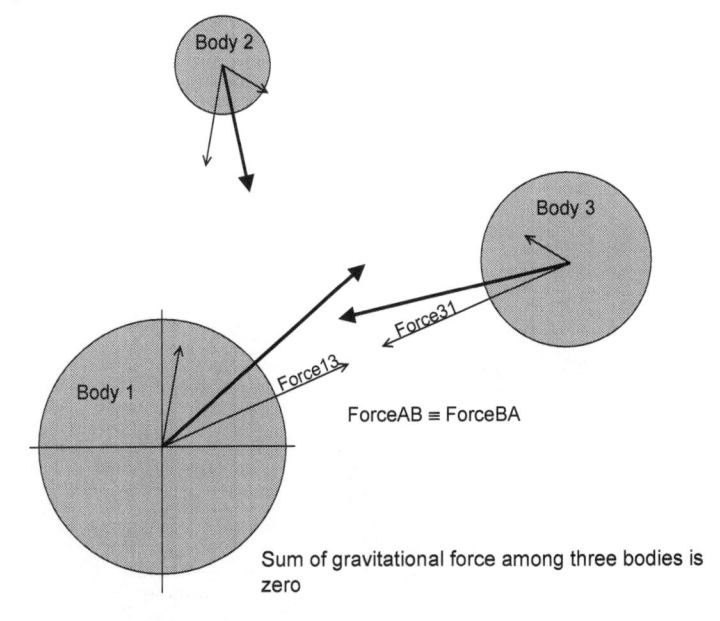

Sum of gravitational force among three bodies is zero

ForceAB ≡ ForceBA

Summed up force vectors produce a net force that is also zero. The force at any individual body is not zero but the sum between any two bodies and among all three bodies is zero. The resulting force vectors in the picture above shows the force at but one instance of time. Forces change magnitude and direction as soon as the distance between bodies changes – which accounts for chaotic movement – but the net force among all three bodies sums up to zero at all times. The only remaining requirement is to have the quantized force acting on all bodies to be of the same time duration. But the gravitational wavefunction is an even function and its reduction is always equally balanced. The force-time product, at times called impulse, is no other than momentum and the momentum created in all bodies adds up to zero across all bodies. Therefore, the conservation of momentum holds.[75] The conservation of momentum must hold for two, three, or a multi-body system because a net increase in momentum would result in, or require, net energy. A three-way link arises only in the sum-zero framework. This mechanism is intrinsic to every even function and if the wavefunction reduces it can do so only in the sum-zero manner.

For existing links, link creation or re-creation happens with interweaving, which is considerably faster than the speed of light. Interweaving can be visualized as merging of links that are alike, and the end effect is that gravitational links that are reaching out merge with identical neighboring links, forming stronger links. The operation of wavefunction superposition is inherently instantaneous. If the interweaving of gravitational links is to be instantaneous, the likeness then refers to the geometric overlap in space. The gravitational wavefunction is two-dimensional and can be visualized as thin sheets; both sheets need to overlay each other in one plane if these were to add instantly. In general, information is about virtual entities and virtual entities are information: alike virtual entities become linked-and-merged instantaneously. The creation of a *new* gravitational link, however, has the limit of lightspeed. The interweaving of virtual gravitational links happens with the hyperbolic function and it is likely that this mathematical relationship is a learned relationship. The hyperbolic relationship is a symmetrical and inversely proportional relationship. In the case of gravitation the inverse relationship is also squared and then the symmetry is in proportion rather than in the magnitude of its numerical value.

A photon's reduction realizes energy at one locality while an electron reduction realizes the mass of the electron at one locality. The

---

[75] To show that impulse is the same as momentum, we start with **F = m·a**. Acceleration **a** is the change in velocity **v/t** and then **F = m·v/t**. Rewriting, **F·t = m·v** where **F·t** is the impulse and **m·v** is the momentum **p** of the body. If all forces **F** sum up to zero and if **t** is of fixed duration then **p** sums up to zero as well and, therefore, the overall momentum in a multi-body system is conserved. Mathematically, if ForceA = ForceB then these forces are equal for some given set of conditions. For gravitation, however, ForceA ≡ ForceB and that means that forces are equal at any and all times and thus forces are equivalent. It is tempting to speculate that the fixed time duration requirement drove Newton's postulate of the absolute time. While the gravitational wavefunction reduces instantly and has no time parameter, the periodic nature of the reduction still needs to be addressed.

reduction of gravitwins (gravitational links) manifests as real momentum at the peripheries of the links at all interlinked bodies in the framework of momentum conservation. Quantum mechanical gravitation is about the distribution of moving energy. All realizations happen in the context of the conservation of energy where the energy parity is conserved. In the absence of the reducing gravitwins the momentum is not realized and a body moves at constant speed in a straight line.

In summary, each of the gravitwin creation and reduction operate under a distinct mechanism. The reduction mechanism is instant and periodic, but because the reduction depends on parameters other than time the reduction of gravitwins can be influenced – for time is always a derived parameter. The creation of new gravitwins is instantaneous under spherical geometry, but becomes progressively delayed under other geometries. For example, with the increase in eccentricity under ellipse geometry the gravitwin reach becomes progressively slower and approaches lightspeed at the limit.

Two Components of Matter

Every subatomic particle of matter has two joined components: the real and virtual component. The virtual component is dedicated to energy while the real component is dedicated to the spatial size aspect in the real domain – that is, mass. By definition and without exception, every virtual entity is double ended. When the real component transits into the virtual domain, its single ended parameter representing the spatial aspect of matter becomes a double-ended computational parameter to properly relate in the virtual domain. Both the real and virtual components can independently acquire any shape in up to three dimensions, yet they remain joined in but one point. When the real component transitions to the virtual domain it remains linked through this point with its corresponding virtual energy component.

Quantum Pythagoreans

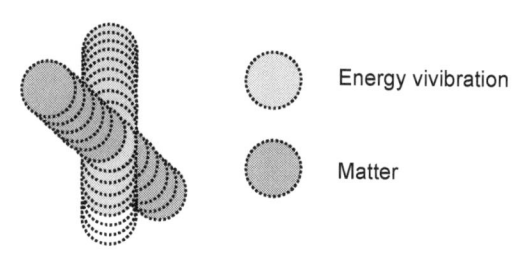

Energy vivibration

Matter

Matter and energy components joined at one point

Coming back – that is, materializing via the wavefunction reduction – the subatomic real component appears in a computable state while the virtual companion remains in the virtual domain but both components continue to be interconnected. Because the now-real and the virtual components have a different computational basis, the interconnection between the real and the virtual components can only be through the zero-dimensional geometry of a point. The two components are "nailed" in one point when the real component is real and having the exclusion properties of the real domain. The subatomic particle must materialize in a point if it is to remain in a computational relationship with its virtual component. Indeed, the real and the virtual components can be visualized as being hinged in a point. As soon as the real component spreads and acquires a virtual representation in the virtual domain, the computability of each component individually exists in more than zero dimensions and the spatial component is no longer confined to a zero dimensional point. Nonetheless, both components remain interconnected in a point because they continue to exist in a mutually computational relationship through this point.

Antimatter

"... or I lay your soul to waste! ... Hmm-yeah" (hoo-hooo)

Mik Jagger
Sympathy For The Devil

Antimatter appears when the real and the virtual components are separated and in consequence the space-energy continuum of a particular atomic particle is broken. The interconnecting "rivet" is now broken and the real component is "stuck" in the real domain. Bringing both of the broken-apart components to the same locality results in matter-antimatter

●●●●○○○ **Gravitation**

annihilation where both components are destroyed – that is, the mutual annihilation transforms matter into energy. The broken-apart entities, when brought together, would now be interacting along one or more dimensions and, not being tractable, mutually annihilate. The resulting photonic energy is twice the energy equivalence of the real component because the real and the virtual components of standard matter carry the same amount of energy. In standard matter only the real component manifests in circumstances such as weight and acceleration. Presently, the matter-energy equivalence is by some believed to be $E = m \cdot c^2$, which ignores both the virtual component's energy and the irreversible nature of the matter-to-energy transformation. The actual formulation is $E \leftarrow 2m \cdot c^2$.

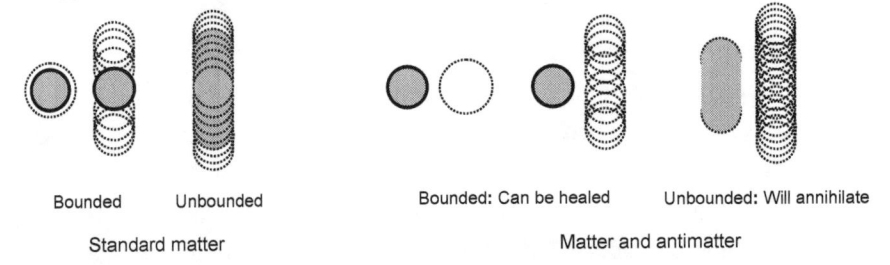

Bounded   Unbounded   Bounded: Can be healed   Unbounded: Will annihilate

Standard matter   Matter and antimatter

Matter and antimatter no longer have the joined spatial (real) and energy (virtual) components

There are no reports of being able to "rivet" or "nail" the components together in a way that would avoid annihilation and restore matter to its space-energy continuum. However, light can bind the *spatial* component of matter into a point, and it is likely that light can be applied to heal matter in this manner. Bounded into a point, the spatial component of matter no longer annihilates with antimatter and the two components can rejoin.[76] A certain amount of energy is needed to break standard matter into matter and antimatter, but after healing this energy is recovered. The photon of light cannot create matter and antimatter because light can impart its own energy only to two real things during reducing transformation – and that is not possible with but one particle of matter such as an electron or a proton.

---

[76] When the virtual electron becomes a real electron, it reappears at one spot because of tractability. The real component – that is, the electron – must be a zero-dimensional entity when it becomes real if it is to remain as standard matter.

Quantum Pythagoreans

Continued accumulation of matter strains computability of the atom because spatial distances diminish and the separations within the atom are necessary for tractability. The tractability limitation manifests by radiation of excess energy and this is the commencement of the solar furnace activity. Tractability within the atom depends on maintaining the balance between the atom's real and virtual energy. Light emission may sustain a particular atom but neighboring atoms are then subject to excess energy. Different atoms have different light absorbing characteristics, and this may favor formation of new atoms when protons and neutrons are available as fusing components. When neighboring atoms close in, electrons cannot materialize in eigenstates they cannot compute and the balance of real and virtual energy fails to hold. The lack of tractability means that an eigenstate is not available and this may result in the separation of the electron's real and virtual components: the real-only electron and the virtual positron. The subsequent annihilation of the real and virtual components conserves energy by converting non-tractable atomic components to energy such that the conservation of energy prevails at the expense of matter. The core's protons and neutrons also maintain tractable states and a considerably larger release of energy can happen if a proton separates from its virtual component. The proton mass is close to 2,000 times greater than the electron mass and the measurement of X-ray emissions can be unambiguously assigned to the electron or proton annihilation.

After the breakup, if the positron does not occupy the same place as the real component from which it was separated, the positron can go on indefinitely; two positrons pass through each other – just as photons – without annihilation.

If there were but a single characteristic that typifies the real domain it is the exclusion property because one location cannot hold two real things. Similarly, inclusion is the central attribute of the virtual domain that allows all virtual entities to exist in superposition. Broken up matter and antimatter are not exclusive to each other. Matter and antimatter are inclusive because the positron is a virtual entity with inclusive property that now interacts uncontrollably with its real-only component. Matter and antimatter will superpose and self-destruct because no entity can ever be both real and virtual. In other words, real and virtual entities are not computable in more than zero dimensions. If the particle is not corrupted and its components are connected in a point, they remain tractable and together can also tractably and reversibly transform. Once broken up, however, matter and antimatter compute in an irreversible transformation. Matter and antimatter will destroy each other not because they exist at the same time but because they overlap in other than zero-dimensional space.

Pythagoreans in their parity transformation of even and odd numbers laid the foundation of mathematical group theory. Atomic entities' wavefunctions can have many shapes but they belong to but one of two groups. The even function group describes the atomic entity that has inclusive property while the odd function group describes an entity that has exclusive property. Even functions are always symmetrical about the vertical axis while the odd functions are always symmetrical about a point. Photon wavefunction, for example, can have the most intricate shapes but mathematically and in actuality will always be an even function that is always symmetrical about the vertical axis while always having inclusive property. The even-ness of the function also explains the even distribution of energy when two atoms or two bodies absorb a photon of light and energy is applied evenly to both bodies. The even-ness symmetry dominates and allows the experimenter to make predictions in cases where the null reading interaction would appear to result in a non-symmetrical wavefunction – the even-ness symmetry prevails and the wavefunction is not only trimmed but reshaped as well to maintain symmetry about the vertical axis. It is apparent that light can remain computational at all times, governed by both the axial symmetry and by energy conservation.

Other elements the likes of electrons and protons must also remain tractable, for tractability guarantees the reversible transformations between the real and virtual domains. If for any reason the transformations encounter computational difficulties antimatter may appear in the real domain. Matter's real component is an odd function while the energy component of matter – now antimatter – is an even function. While sporadic outbursts of x-ray photons may point to localized annihilation of matter into energy, the event of a nova or supernova reveals a general lack of tractability and consequent large-scale synchronous or avalanched and irreversible matter-to-energy conversion. A sun going nova will quickly lose matter by such nonreversible transformation, for it had contained mass not capable of remaining tractable and, therefore, organized.

I can become nova
Destroyer of suns

I can become brother against brother
Destroyer of civility

I can become surreal
Destroyer of self

The pulsar flips the truth to false, false to true
Round around eternity
Feel the concussion
                 pretty dust novae
Brain dead for posterity

The Source of Tractability

One, two, three, four – cannot count dimensions four
One, two, three – thanks for making house for free

How does one arrive at a tractability rule? How could a very large number of bodies be or become tractable? What should be the proper mass ratio of the sun and the planets? How far apart should planets be? The only way to resolve these issues is to take the elements essential to tractability and have them compete with each other for their own preferred way of achieving tractable solutions. Two-body topology has a family of solutions and so does a one-body arrangement because the singular mass-accumulating sun is not the only way of achieving single body topology. Galaxies have one-body topologies as well because all solar systems within the galaxy move and rotate in lockstep as one body. Galactic patterns, however, are the results of different computable mechanisms that work solar systems in different though computable ways.

The interim but not necessarily easier answer is that there can exist any locally tractable topology but any growth in such topology must also be tractable as its context enlarges or changes. Starting at a sun with its one-body topology, the expanded context moves to dual sun topology or to sun with single planet two-body topology. Further context enlargement, while maintaining tractability, results in multi body but orbit-interlocked planets forming solar systems like that of our own. Additional expansion results in one-body galactic topology but with various computable patterns. Explanations remain for galactic clustering, planetary rings, and specific computable contexts for various and glorious galactic shapes.

Pythagoras through his Tetractys first alluded to the universal order. He used a triangle with ten dots inside it to represent the universal order. Since the visualization of computability is similar to the Pythagorean Tetractys, it is worthwhile to build on that and a triangle is a good start when working with three variables. A more general interpretation is coming up.

Pursuing computability a bit deeper, the realization that the atom cannot even exist without being tractable at all times points to atomic tractability as the source of all computability on the macro, or cosmic, scale. This is really not a philosophical sidetrack because the gravitation-neutralizing spaceship has atomic computability at its engineered heart.

Two readily identifiable parameters essential to tractability are mass and distance. The third parameter is force. If each parameter of force, mass and distance compete for a maximum of some unit, then the competition among these three parameters results in each parameter wanting to take the full measure of such unit. But even if one parameter claims the whole unit, the remaining parameters would never be completely zeroed out. Force, mass and distance may each be dominant but force cannot exist without mass and distance, for example. Integers and near-integers appear to have a role to play in the solutions, and hyperbolic functions may just be the ones that could describe the tradeoffs. The force parameter is about generic force and can be gravitational or electromagnetic or whatever. The ordinary experience of the gravitational force is that of attraction but gravitational force also produces torque. Torque results in angular momentum – it is force that manifests by rotation and orbits.

A four-dimensional or 4D world is treated theoretically on paper and has certain characteristics. One of the characteristics stands out because 4D objects were shown to have non-computable surfaces. Even if an assumption is made that a couple of 4D surfaces are topologically identical or topologically equivalent, this cannot be mathematically proven. In three dimensions a pillow, a ball, and a cube are topologically equivalent and the equivalence can be mathematically proven because all three have common and repeatable and unique characteristics. Also, a mathematical proof can be had in 3D where a wheel rim and a hubcap are not topologically equivalent. However, there exists no computational solution for determining if any two 4D surfaces are or are not the same. There exists no repeatability in 4D and four or more-dimensional worlds cannot contain computable systems that would yield a yes-or-no answer, for example. Repeatability is a property where the *if-then* inference can be constructed and can be verified to hold, but in the 4D world the *if-then* construct does not exist simply because it cannot be constructed to hold in 4D – that is, the *if-then* construct is not enforceable in a system with four or more degrees of independence. Even though the movement of bodies is at times not tractable in 3D if bodies become chaotic, and even though some tiling problems are not computable in 2D, the lack of computability is the norm in the 4D world.

Whatever elements, whatever conditions, and whatever forces hold matter together, there is no possibility for matter to exist as real matter while taking four degrees of independence.

Therefore, the three variables of force, mass and distance each compete for no more than three degrees of independence. The force, mass and distance each compete for a full measure of three degrees of independence but do not compete for more than three degrees of independence. The force, mass and distance can take the full measure individually. Yet it is also apparent that the force, mass and distance cannot take more than the full measure collectively. The variables of force, mass and distance acquire three degrees of independence and the three degrees cannot be exceeded by the three variables *individually or in summation*. 'Individually or in summation is three' is the Maxim of the real domain.

●●●●○○○ Gravitation

Getting The Move On

More the wing you spread
Weave the waves ahead
Friendly crest
    steers the best
Friendly hail
    trim the sail

A sputtering sun convulsing at times with matter-antimatter annihilation was all there was to the very early universe. New, more complex atoms were created inside the sun and a greater variety of atomic creations likely produced new computable knowledge. As much as matter can and does accumulate into a one-body system while remaining computable, atomic stability puts limits on such growth. The early limiting mechanism is the firing up of the solar furnace while the nova mechanism produces a more radical adjustment. However, if two bodies could orbit each other then a great advance in the organization and expansion of the universe begins, for orbiting systems allow even more matter to appear – and the universe can then expand beyond the single body of a single sun.

Other single suns also began to grow and glow while separated by great distances. Bodies move when force is applied. A single body growing in size is extending its gravitational reach and at some point it begins to accelerate toward the nearest and largest body or cluster of bodies. The idea now is to get the moving bodies organized by orbits or by creating other bodies to form computable topologies.

There are a maximum of three degrees of independence in the real domain. In all of the earth's history all references to degrees of independence were always occupied by spatial distance of open space. It makes much sense, then, to start with distance. In general, however, force and mass are also competing for the three degrees of independence. All three parameters compete for independence because force, mass, and distance each can produce a computable solution if each of them dominates. Spatial distance that can freely advance is also closely associated with the word 'dimension' and so presently the mental task is to allow force and mass to have their dimensions of independence as well. A degree of independence and a dimension of independence can be used interchangeably.

Spatial distance is usually not independent in all three dimensions. Sharing the three independent dimensions among force, mass, and distance should not be a difficult concept. On the planet's surface, for example, things stay where they are placed and do not move freely or independently in all three dimensions of spatial distance. Therefore, not all three dimensions are spatially independent on the surface of the planet.

Spatial Distance Takes All

In this case spatial distance has full dominance. A body far away from other bodies goes on predictably – that is computably – on its way because forces are nil and there is no other mass in the vicinity that is considerably larger. Spatial distance (or just distance) dominates because distance can be anything it wants.

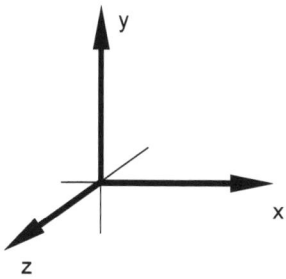

Three independent dimensions of spatial distance

If the three dimensions of distance are to be independent of each other, the independent movement is orthogonal to each distance dimension. A body moving computably in a context where all three degrees of independence are taken by distance is simple; it just keeps on going in a straight line.

Each computable context is described with the count of degrees of independence or dimensions of independence. In this case the spatial distance has almost three, mass has near zero and force also has near zero dimensions of independence. Mass has near zero dimensions not because it is small but because there are no other bodies in the vicinity. Similarly, force also has near zero dimensions because the distance between bodies is so large the force is nil. Using the convention of starting with force, then mass, and finally distance, the context where distance takes all dimensions of independence is 0,0,3. Each possible computable context is the *hyperstate*. Although force and mass are shown as having zero degrees of independence, both force and mass are close to zero but are never zero. Claims by distance to degrees of independence are close to three and this can be called the near-integer. One could also show this hyperstate as 0+, 0+, and 3−. All degrees of independence always add up to exactly three.

Hyperstate 0,0,3 can be called empty space. Newton's first law of motion covers the computability of hyperstate 0,0,3. A moving body in this hyperstate is computable; by measuring its velocity at some point, a single straight continuing path of the body is computable. Newton did not explicitly call his laws computable, but a search for the equation is a search for computability. Newton was acutely aware of the absence of an equation governing three bodies in general. Much credit should go to Newton for discussing the subject even though he did not have ready-made answers at that time.

Force Muscles In

When two bodies interact in its simplest form, they do it in one dimension. One-dimensional interaction is along a straight line between the two bodies.

When the force of gravity arises between two bodies, force exists in one dimension (1D) along a line connecting the two bodies. The force of gravity attracts two mass bodies to each other, and, along this dimension the two bodies no longer behave as independent bodies. Force can now change substantially while distance changes little. Force begins to be dominant and distance cannot be "anything it wants." Distance has lost its independence in one dimension. Bodies are still far enough apart from each other for them to appear to each other as dots. The bodies' mass is, therefore, near zero-dimensional and in such context the mass continues to have no claim on dimensions of independence. Mathematically, force and distance relate to each other through a function which, when drawn, is a hyperbole and this solution means that the two bodies are computable. When bodies are far apart, distance is independent because the force is nil. When bodies move closer and the gravitational force grows, the dependence-independence reverses and it is the force that is now independent or dominant. Force has taken one independent dimension away from spatial distance.

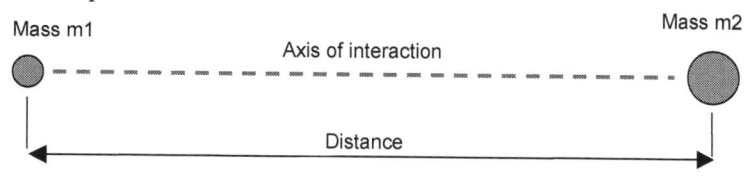

```
Mass m1                                              Mass m2
              Axis of interaction
  ●─ ── ── ── ── ── ── ── ── ── ── ── ── ── ●
  |                                                  |
  |←──────────────── Distance ─────────────────→|
```

Force claims one independent dimension

This is hyperstate 1,0,2. Spatial distance continues to hold two degrees of independence because bodies can move freely in a 2D plane that is perpendicular to the axis of interaction. Even though the gravitational force is as prescribed by the separating distance, both bodies are free to move in any direction of the perpendicular 2D plane. When distance diminishes further, such perpendicular movement can result in bodies orbiting each other. Once bodies orbit each other the hyperstate 1,0,2 is maintained and the parameter of periodicity appears along with periodic force. Of significance is that bodies approaching and then orbiting each other are computable at all times, either as approaching bodies or as orbiting bodies. Of note is the realization that two orbiting bodies continue to have force acting in one dimension: The force acts along the axis and the bodies may be orbiting each other, but the force acts in one dimension because it always acts along the line of the axis.

Hyperstate 1,0,2 covers all two-body interactions where the formulation is in several though related equations – for both the linear and angular movement. The work here is by Kepler, Galileo, and Newton.

You can survey the independence by tugging at some variables. Flattening the shape of the bodies in a plane perpendicular to the axis does not infringe on the independence of either the distance or force in this context overall. Locally, a gaseous or a pliable body would not resist the flattening as a solid body would. Also, a body spin would not challenge the independence of force along the axis of interaction and bodies are free to rotate without affecting the force between bodies. Further, any spin will have a tendency to flatten the body. A single solid body would not flatten but a single-body configuration of a galaxy can and does exist as a flat unified body and it can be created flat right from the start if large angular momentum is called for. A new definition of inertia fits well at this point and inertia is addressed in the upcoming chapter. Afterward, additional hyperstates are forthcoming.

New Definition of Inertia

With his rolling balls experiments Galileo concluded that in the absence of friction a moving body could move on forever and thus a body's momentum is conserved. Since the conservation of momentum can be formulated in terms of the conservation of energy, additional energy is required to increase a body's momentum. Similarly, a moving body will produce energy if its momentum decreases. Newton's law of inertia states that a moving body will resist a change to its direction or speed where the change in direction or speed is the change in the body's momentum. Newton quantitatively addressed Kepler's proposition that inertia is the property of mass.

Inertia is the attribute of mass that arbitrates the change in linear or angular momentum of mass. Mass body at rest will resist force to get it moving but, even if the body is moving, inertia will also resist the change to slow it down, speed it up further, or change a body's direction of movement. Inertia, then, can well be seen as the enforcement agent of the conservation of real energy that is the moving energy. Once energy is applied to get a body moving, the same amount of energy can be recovered when bringing the body to rest. Because the conservation of energy holds everywhere in the universe, we can extend that to say that inertia is the same anywhere in the universe. If a body receives a certain amount of moving energy and travels to another part of the universe then the same energy must be recoverable someplace else only if inertia remains the same. Inertia thus facilitates the storage of energy as a moving real energy and, in reverse, inertia facilitates the recovery of energy as well. Inertia not only stores the energy but also stores the direction in which energy is applied. The storage and recovery of energy *and* direction is the property of inertia. Inertia that is associated with linear momentum is proportional to mass, as a more massive object requires a proportionately greater force to bring it to a particular speed. Angular momentum, though, is a function of not only mass but also of the geometry under rotation. Two bodies that weigh the same have the same mass but, while spinning at the same rate, one body will have different angular momentum if bodies have different geometries. Mass further away from the center of rotation will have greater angular inertia because the speed along the periphery of the circle is greater farther away from the center.

Newton formulated inertia from the perspective of the resistance to change. Inertia can also be taken as a mechanism of energy conservation, which is the same as the mechanism of storage of real work. Inertia's mechanism is reversible in that one form of real energy such as linear work is transformed to another form of real energy such as potential energy. The actual mechanism of the inertia deals with the transformation of the energy between its real and its virtual forms. A moving body acquires vivibration that is proportional to the energy imparted onto the object in absolute terms.

There is more to inertia, for inertia also has the potentia to maximize momentum and, therefore, maximize storage of moving energy in a local structure. For example, if a body can commence rotating when subject to gravitational force, such body then resists the increase in velocity even further than with the linear motion alone. Newton's take on inertia is now enhanced in that a mass body will resist change to the *maximum computable extent*. If the gravitational force acts along the direction of travel and the body is accelerating, the new formulation of inertia allows the body to commence spinning and thus further resist the increase in speed. Moreover, the new inertia framework enables a body to commence spinning and create a flattening disk structure, *both* of which maximize locally stored moving energy. An object that flattens acquires greater angular momentum for a given rotation and thus offers even greater resistance to the applied force. In summary, (**1**) inertia is the enforcement agent of the conservation of real (kinetic or potential) energy and (**2**) inertia facilitates the greatest possible real energy storage by creating and maximizing angular momentum.

One outcome is that the angular momentum of a spiral or a bar galaxy points in the direction of the galaxy's linear movement.[77] Keeping

---

[77] Bending the fingers of your right hand in the direction of the spin, the thumb points in the direction of the angular momentum. Additionally, the

up with the expanded definition of inertia and, given that every solar system in a galaxy is moving toward the center of the galaxy, the greatest or the primary angular momentum of every solar system in a galaxy points to the center of the galaxy.

Mass body subject to gravitational force seeks to acquire the computable linear and angular momentum to achieve maximum overall momentum. The body or bodies within the systems rotate and flatten out to achieve the maximum angular momentum. The classical equation relating force, mass, and acceleration, or **F = m·a**, does not consider a change in spin and is, therefore, in need of updating. For most earthly activities the update is not necessary and indeed a ball falling from the tower at Pisa does not spin appreciably when it reaches the ground. Yet, any increase in spin is subtracted from the linear speed. For other than gravitational forces – such as mechanical force – the spin will not develop if only the linear component of force is present. With the new definition of inertia, the overall effect is that a considerably greater gravitational force is required to linearly speed up or slow down a system that can maximize its angular momentum while remaining computable.

A body subject to net gravitational force will resist an increase (decrease) in its linear velocity by increasing (decreasing) its angular velocity and by increasing (decreasing) the body's angular inertia. A solid body may acquire rotation (spin) but will not change its angular inertia much as it cannot flatten very much. However, a computable system under acceleration such as the solar system or the galaxy will. This is the primary mechanism that aligns all planets in one plane while orbiting the sun in one direction, for this arrangement maximizes the angular momentum of a system under acceleration. This leads to the proposition that a galaxy that is not subject to acceleration will likely become a spherical galaxy. The constituent solar systems of a spherical galaxy have symmetrically and radially oscillating, rather than orbiting, dynamics.

Inertia mediates the conservation of real energy. The underlying mechanism of inertia then requires a link to the virtual domain to explain

---

plane of the galaxy is perpendicular to the acceleration and the absolute time reference may then hold across the entire galaxy.

the computational component of the enforcement. As a physical law, the conservation of energy is easy to postulate yet it is somewhat more difficult to explain how the overall conservation mechanism operates. One necessary component of the mechanism is an intrinsic computational parameter that is readily available and that indicates a moving body's quantity and direction of energy. The linear momentum of a body has equivalent vivibration. Angular momentum also has equivalent vivibration in the form of the virtual rotational frequency. The sum of the linear and angular virtual frequencies then computationally represents the total moving energy of a body. A body's linear and angular vivibrations are computational parameters in the virtual domain that enforce the conservation of energy. An important aspect is that vivibration is absolute – that is, each moving body has its own particular linear and angular wavelength that is a measure of its moving energy in absolute terms. A moving body's vivibrations, both linear and angular, are then applied as computational parameters. If force arises to slow the body down and decreases the body's momentum then such force must procure actual work because the decrease (recalculation) of the body's vivibrations gives rise to real momentum that is overcome by the applied force. In reverse, the increase in real momentum increases vivibrations. Inertia is at the center of the two-way (reversible) energy transformation mechanism where real energy speeding up an object is transformed and saved as virtual energy, which in the form of vivibrations is ready to transform back to real energy.

The most intriguing aspect is the bypass of inertia's transformation mechanism altogether. Virtual energy can be directed at a moving body in such a way as to computationally interact with vivibrations of the body and actually increase or decrease the frequencies of vivibration. There is, then, the ability to stop moving bodies at a distance. The conservation of energy continues to hold because the virtual energy must be expanded in the interaction with the body's vivibrations that results in a change in the body's moving energy.

The encouraging part is the use of vivibrations to explain the conservation of energy at speeds near the speed of light. At body speeds approaching the speed of light the energy enforcement needs to deal with the vivibration's computational speed being possibly limited by the speed of light. A body approaching the speed of light can acquire energy without bound and, even though the actual speed of the body does not exceed the speed of light, vivibrations can and do increase without bound. Technically, infinite vivibrations equate with infinite virtual energy. Although the speed of a real body never reaches the speed of light, the energy continues to be conserved because the energy that is imparted on the body is accepted and can be later recovered. A body at near the speed of light increases its vivibration without bound – yet the body's mass does not increase. Even at light-limiting speed the mechanism of the conservation of energy is in agreement with the vivibration computational mechanism. The outcome from this is that a body at light-limiting speeds has no need to increase its own mass that would be different from its mass at rest, for the conservation of energy is satisfied. No moving body exerts a gravitational force on other bodies that is greater than the body at rest. Inertia's ability to mediate the conservation of real energy is unbounded.

In a laboratory experiment, an increase in an unstable particle's speed is correlated with the delay in the particle's breakup. While the half-life of an unstable particle increases with increased speed, the dependent variable of time reflects the increase in vivibration and thus the particle is stable for a longer period because its own vivibrations increase. The increase in a particle's energy increases the half-time-of-decay parameter. A mass particle at lightspeed does not decay because it contains an infinite amount of energy.

> *Screaming compendium*
>   *Quiets down, slows down, melts down*
> *But will not blow*
>   *Your head off*
> *If you think*
>   *While you have*
>   *Your head on*

Meanwhile, the reality wardens claim that a clock traveling at some speed on an airplane slows down measurably. That may be the case but it is the atomic clock with its half time of decay property that is the issue here because the half time decay parameter is a function of absolute speed. The clock that is not based on atomic decay but, instead, uses a light-based clocking device will not change its time as it moves and is indeed the absolute clock.

Force Takes Another One

A smaller body that is in orbit around a larger body moves in a near circular or elliptical path where the orbit path can be visualized as being in a plane. When another body joins in the same orbiting plane and has another independent orbit around the larger body, the force now has the ability to act in any *two* dimensions in a plane and the compound two-body arrangement becomes another hyperstate. The radial forces issuing from the large body can now move independently and, therefore, can act anywhere in the orbit plane. Body separation continues to be substantial enough so that orbiting masses have near-zero degrees of independence. Distance can have up to one remaining degree of independence if this arrangement is to continue to be computable and the independent distance dimension is perpendicular to the force dimension. This allows the orbiting plane, along with all orbiting bodies, to acquire a tilt while remaining computable. Addition of more bodies in the orbiting plane adds more lines of force but if all lines are within the orbiting plane then the force does not claim more than two degrees of independence.

Orbital interlock introduces an interesting aspect to force independence. Forces extend radially along a plane but the movement of such force lines is not independent because there is a mathematical and deterministic relationship among the angular movement of lines of force. Orbiting bodies in the orbital interlock merge into a single body and force retains but one degree of independence, which corresponds to an earlier hyperstate. For force to claim two degrees of independence, the orbiting planets would each have "their own" elliptical orbit.

If three or more planets were not all orbiting in one plane then force takes the third and the last degree of independence because the force lines could act anywhere along three dimensions. If a computable solution were to happen, both the mass and distance would have near zero degrees of independence and all bodies would then be subject to some computable – though not necessarily lockstep – movement. If distance loses the last degree of independence, none of the orbits would be independent. In other words, orbiting bodies could computationally merge into a unified body but the resulting body would not be free to change its orbit distance or inclination, for the distance dimension has lost all degrees of independence.

●●●●○○○ **Gravitation**

This is technically workable and this particular hyperstate is discussed further on.

A single orbiting body has the freedom to move in any orbiting plane because the line of force always takes but one dimension. A moon's orbit plane and a planet's orbit plane can point in different directions if the moon's angular momentum is needed for something else. Multiple orbiting bodies also have the freedom to be in different orbiting planes but the hyperstate described here has all bodies orbiting in one plane because in the present case force takes no more than two degrees of independence.

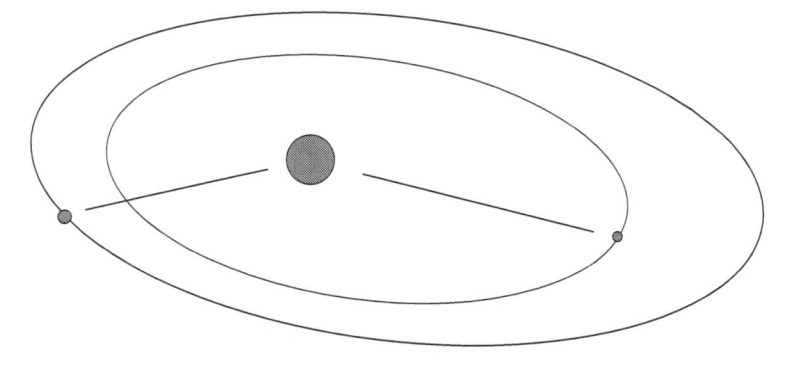

Computable context of two bodies with force having two dimensions of independence

This is hyperstate 2,0,1. Kepler obtained the mathematical solution of this hyperstate and the orbits are called Keplerian orbits. Although Kepler's solution does not allude to all planets being in one plane, this is indeed the requirement for computability of a compound two-body system. Each planet individually deals with the sun as a two-body subsystem but the sun has forces going toward a plurality of planets in but two independent dimensions if the planetary orbit times are not interlocked. When all planets are in one plane, all of forces' directions are along the plane and force does not exceed two degrees of independence. If all of the orbiting bodies interlock in their orbits, however, the orbiting bodies become effectively a single body and the hyperstate is no longer 2,0,1 as all merged bodies produce just one computable line of force at all times – that is, the variable of force goes from two degrees of independence of a plane to one degree of independence of a line. Note that the merging of, say, two lines of force always happens through linear superposition, but the idea of reducing 2D force into 1D force is that the merged 1D force line is tractable after it is computationally merged.

Quantum Pythagoreans

In the planet-only context the hyperstates of Mercury, Earth, and Mars could all be different because Mercury has no moons, Earth has one, and Mars two. Planets having three and four moons can be in two additional and different hyperstates as well. Taking Oss as a whole, however, all planets merge with their moons and form one hyperstate because planets and their moons merge into near zero dimensional mass objects. Hyperstates change as a result of the observational context but all hyperstates represent a computable solution. Venus and Earth's orbits are interlocked and this also leads to an observation that hyperstates are not exclusive. Venus and Earth can be seen as a hyperstate with two independent orbits where force has two dimensions – or a hyperstate with one (merged) orbit with force having but one dimension. A particular geometric arrangement can belong to more than one hyperstate and also a different (larger or smaller) context results in a different hyperstate. Indeed, this is where 'hyper' in hyperstates comes from. As one increases the distance from Earth on out, you encounter several hyperstates. Furthermore, the change in hyperstate is not by distance alone. Moving out from Earth and including the moon, the next body is the Sun because the Sun is the most dominant body after the moon. After the Sun, both Venus and Mars come in as the Sun-Venus-Earth trio, as well as the Sun-Mars-Earth trio, each form another hyperstate – effectively a Sun and one-body hyperstate. Here is where the Earth-related hyperstates end, for the inclusion of other bodies does not involve Earth. Other planets can then be included as components of the Oss.

Venus-Earth is in effect a single planet because of their orbital interlock. One can calculate the equivalent or effective Venus-Earth body, which, having a highly elliptical orbit may even spend some of its time inside the Sun. Ideally, the gravitational center of all planets would be at the center of the Sun at all times and in such case the Sun would be in dynamic balance in relation to all combined planets. This is mathematically expressed as being in a minimum about the 'second moment,' and is no other than having your automobile wheel properly and dynamically balanced. The equivalent single Venus-Earth body averages out the disturbance of both planets. The orbital interlock having the ratio from the musical octave then lowers the Sun's wobble. Because the Venus-Earth planetary combo can be tractably merged into a single body, one should not have difficulty visualizing one of the planets as "missing," for this is a good example of the reduction in disturbance due to the second moment. Should Venus and Earth be physically merged, the orbit having an identical angular momentum would create a much greater dynamic disturbance on the Sun. Planetary orbit interlock then results in angular momentum with a smaller disturbance onto the Sun. Orbital interlock also exists between other planets as well. By computationally merging all planes into one, the overall movement or wobble of the sun can be calculated. The magnitude of the wobble increases and decreases during the overall period and such excursions or harmonics distribute or "spread out" the greatest magnitude of the wobble. Should you observe the wobble at another solar system, you would be able to calculate the overall merged body that creates this wobble. However, it is not possible to reconstruct the full complement of planets in that solar system, for many different planetary arrangements can result in an identical wobble profile.

Mass Weighs In

As three or more mass bodies begin to come closer they are no longer mere zero dimensional points. Mathematically, a mass of bodies is known, distances are known and, therefore, all forces are known. But there is no function that could be used to calculate the trajectory of any of these three bodies. Newton sought a formula, any formula, which would determine the path these three bodies could take. Having the solution in hand, one could calculate and draw the path in advance and declare the problem solved. The difficulty here is that when the third body comes closer and begins to exert gravitational force in a new dimension, the two-body system changes into a three-body system that does not have a general solution. Without a solution the paths of all bodies become unpredictable.

Quantum Pythagoreans

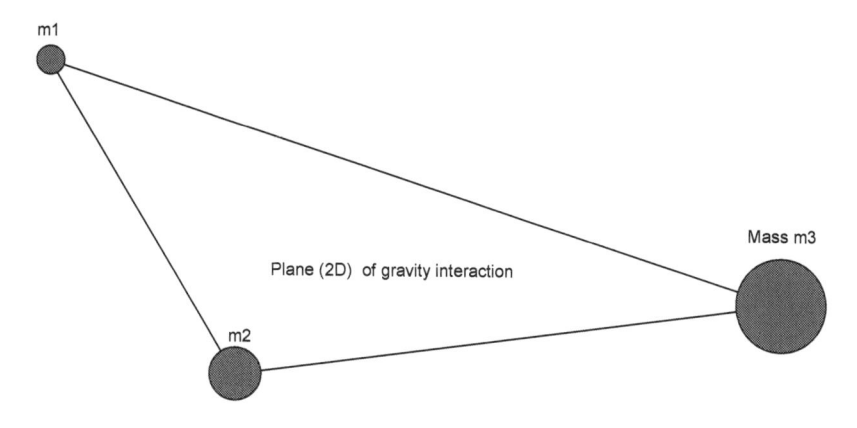

Chaotic interactions of three bodies with force taking two dimensions

Putting three bodies in the hyperstates framework, force takes two degrees of independence because between three bodies the force now exists in a 2D plane. Each body has one degree of distance independence that is perpendicular to the 2D force plane because each body can move freely in that direction. This brings up the total count to a sum of three degrees of independence. However, mass is now closer and no longer a zero dimensional point. Mass begins to claim an independent dimension for itself. In the hyperstates framework the three-body setup is not computable because the total number of degrees of independence exceeds three. Adding more bodies outside of the plane, force continues to increase its claim because force can then act in 3D. The limit on three degrees of independence is exceeded and computability in general ceases to exist for a system in excess of two bodies.

Rocks, trees, and people are indeed three-dimensional. On a larger scale a planet can be so small it makes a zero dimensional point. The working framework of hyperstates is that the absolute size is irrelevant but a dimensional size or dimensional claim is relevant. Standing on the surface of the earth, hyperstates framework establishes the earth mass as occupying two dimensions because the solid earth forms a 2D plane as far as one could see. While it is easy to agree on a context when force acts in one dimension or when the surface of a large body occupies two dimensions, it may not be clear at which point a single round mass object could claim one full degree of independence. This issue does not come up in two body interactions because two bodies are always computable in sub-orbiting (parabola), straight (falling or colliding), orbiting (ellipse or circle), or flyby (hyperbole) configurations. The computability of two bodies is maintained because in all two-body geometries force exists along but one line connecting the two and force can never claim more than one degree of independence. Even if two bodies are very close and each fills up a substantial solid angle in a view, each body can nonetheless be substituted with a point.

Fundamentally, if a body can be orbited then its mass is near zero-dimensional. If a body can be orbited in one direction but not in another direction then such body has one-dimensional mass. If a mass body cannot be orbited then mass is two-dimensional. The dimensionality of mass is a function of collision avoidance – that is, if a body can be orbited then a collision cannot happen and mass in effect becomes zero-dimensional. The hyperstate solution, then, is also a function of the orbiting dynamics. Treating mass body as a zero dimensional point was initiated by Newton and continues to be applicable with the orbiting constraint.

In a three body setup force takes two independent dimensions and force can now be "anything it wants" along a 2D plane. There is then only one independent dimension available if computability is to hold. This one computable outcome results in the only hyperstate where dimensions of independence are not integers or near-integers. We will return to it because the dimensions of independence are allocated in a ratio and results in the topology of our own spiral galaxy.

Another solution is to compose three or more bodies into a straight line. The success of having bodies like beads on a string is that force acts along one dimension; the second independent dimension is taken by mass and the last dimension is for spatial distance that allows bodies to move perpendicularly to both force and mass in any one direction.

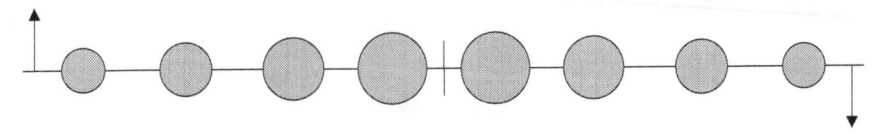

This is hyperstate 1,1,1 and manifests as the barred galaxy. If force is not to exceed one dimension of independence, the entire lineup must rotate in lockstep. If you look at the barred galaxy from some distance the mass of bodies indeed forms a one-dimensional structure or one-dimensional stick. The rotation of the entire bar galaxy slows down the

Quantum Pythagoreans

forward movement of the bodies but bodies remain in a straight line. Mass continues to hold 1D of independence and mass can appear and grow along the bar because the computable topology exists along the 1D bar.

2D Is Flat

The surface of a large mass body has two dimensions of independence because mass is "anything it wants" in 2D. Force takes the one remaining dimension.

Distance is left with none and, indeed, an object not so strangely falls toward the surface and, upon touching the surface, reaches this hyperstate. If distance were to retain one dimension of independence then a body could move on forever in any but one direction, which is not the case at the surface of a planet. On top of the surface the object is not free to move and comes to rest rather than proceeding forever on a straight path.

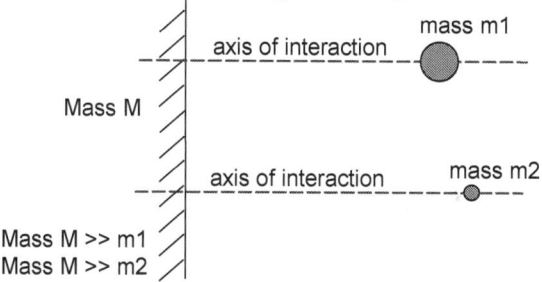

Mass claiming two independent dimensions while force claiming one

This is hyperstate 1,2,0. Force between a large object such as a planet and either of the two smaller objects is much greater than the force between two smaller objects, and thus force is independent in but one dimension along the axis of interaction. Objects rest at the surface subject to one-dimensional force and that is the beginning and the end of this hyperstate.[78]

---

[78] This may seem trivial and perhaps it is. Yet, all mathematical sets have unitary operators. Operation by the unitary operator results in no change at all. This is analogous to multiplying by one. Then again, sitting on a beach and enjoying the view can be quite a blissful hyperstate.

233

●●●●○○○ **Gravitation**

Each small object is independently computable as it moves toward the surface, and before the object touches the surface the hyperstate is that of the two-body system that is 1,0,2. The collision with the surface changes hyperstate 1,0,2 into 1,2,0. The collision invokes the dimensionality of mass because the collision could not be avoided. Prior to collision, all objects accelerate toward a 2D mass at the same rate and this can be logically corroborated as follows:

Two objects having the same weight are falling at the same rate. Because neither object gets ahead of the other these two objects can be joined together and even though the object is now twice the weight the resulting object is also falling at the same rate.

This powerful logical proof precedes Galileo; Giovanni Benedetti first proposed that all objects subject to earth's gravity are falling at the same rate.

3D Is Sun

For mass to claim all three degrees of independence it just keeps on accumulating. The mechanism is for matter to fill everything with itself and matter then claims to be everywhere in 3D, displacing distance and force. It is apparent that inside a mass body there is no independent movement along any spatial distance. Inside a solid body, force is fixed as well and cannot claim to be "anything it wants." This is hyperstate 0,3,0 of a large single and generally solid body.

Although a galaxy moves as one body, on a finer scale the mass does not fill in "every space available" and within the galaxy there are regions consisting of many other hyperstates. Mass, then, can also limit its accumulation that results in computable geometries afforded by hyperstates.

Mapping Hyperstates

There are three variables or parameters competing for the maximum of three degrees of independence. The three variables of force, mass, and distance take on near-integer values of degrees of independence and always add up to exactly three, which produces the computable context of hyperstates. While the source of force may be different on the cosmic and atomic scales, a force that attracts or repels is conceptually the same force parameter. Hyperstates are mapped onto a triangular plane as shown below.

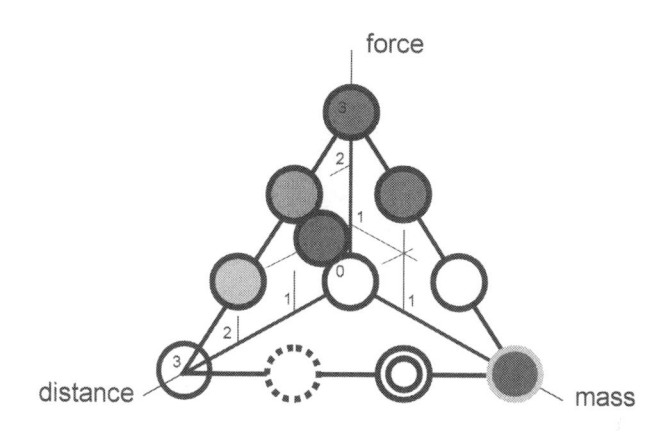

Starting at origin where all degrees of independence issue, the force, mass, and distance each increase along their own axis and in general relate as follows:

$$f + m + d \equiv 3$$

Formally described, this equation shows **f, m,** and **d** representing the degrees of independence taken by force, mass and distance, respectively, while their values are any real number greater than zero and less than three. Degrees of independence taken by force, mass, and distance *always* add up to exactly three and this is indicated with the equivalence sign ≡. This equation defines computability on a macro scale. By giving any variable **f, m,** and **d** integer values of 0, 1, 2, or 3, ten possible hyperstates are formed using this equation as shown below.

● ● ● ● ○ ○ ○   Gravitation

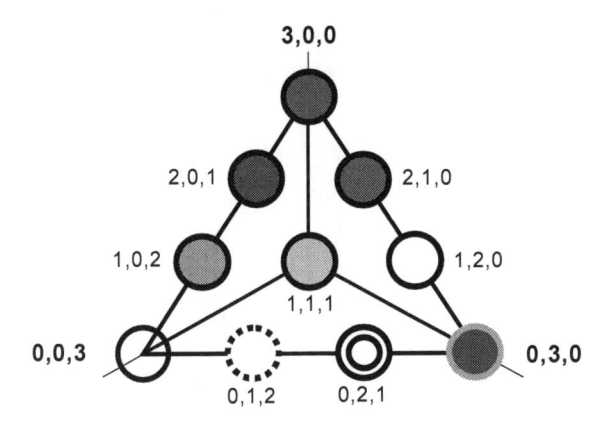

```
                    3,0,0
                      ●
            2,0,1         2,1,0
              ●             ●
      1,0,2       1,1,1       1,2,0
        ●           ●           ○
  0,0,3                              0,3,0
    ○       ○         ◎         ●
          0,1,2     0,2,1
```

If force takes two degrees of independence and distance one, the hyperstate 2,0,1 results; that is Kepler's gravitational orbit law of a compound two-body system of a solar system.

If force takes one degree of independence and distance two, the hyperstate 1,0,2 results; that is Newton's gravitational force law between two bodies. This hyperstate covers two bodies interacting along a straight line or orbiting each other because the force is independent in one dimension in either case. Since bodies with interlocked orbits are in effect a single body, this hyperstate also covers sun and merged planets such as Venus-Earth.

Each of the three variables taking one full degree of independence results in a bar galaxy. This is hyperstate 1,1,1.

Hyperstate 0,0,3 is open space that allows all bodies, including vastly spaced galactic superclusters, to move independently in any straight line.

Hyperstate 0,3,0 is a one-body hyperstate that has the sun at its size limit.

Hyperstate 3,0,0 is a multi-body arrangement in which all bodies move synchronously and no body has independent movement. This hyperstate is manifested as a spherical galaxy. This hyperstate also happens when or if all (moons or planetary) orbits become interlocked while not being confined to orbits inside one plane.

Hyperstate 1,2,0 deals with bodies at the surface of a large body (earth-bound). Force per unit mass is constant here and the constant is called **g** for gravitation.

Hyperstate 2,1,0 happens when a body orbits a bar galaxy in a plane that is perpendicular to the bar. Dual suns also form one-dimensional mass if a body orbits in between them and in a plane that is perpendicular to the line connecting both suns. 1D mass of bar galaxyies or dual suns rotate in one plane and the orbiting body then rotates in a perpendicular plane. The angular momentum of the bar and the angular momentum of the orbiting body can never add up to point in the same direction. Further, an orbiting body introduces wobble to the angular momentum of the 1D mass bar while transcribing a spherical path and the orbiting body also cannot have its angular momentum independent enough for it to become fixed. This hyperstate is likely not very common. Yet this hyperstate would make a fascinating sight. A body would remain between a planet and a sun and viewed from the planet such body would transcribe circles on or around the disk of the sun.

Force and Mass Take It In A Ratio

There is one hyperstate that is not composed of integer or near-integer values. This hyperstate is that of our own spiral galaxy having $1\frac{1}{2}, \frac{1}{2}, 1$ degrees of independence. The primary angular momentum of every solar system in any galaxy points to the center of the galaxy. The intended path of travel for all solar systems is toward the center of the galaxy. The primary angular momentum is the largest angular momentum and the solar plane forms it. The secondary angular momentum is the second largest angular momentum and the sum of all planetary angular moments point along a tangent of the galactic arm. The movement of the solar system is toward the center of the galaxy but most of the actual travel is along the arm. The secondary or planetary angular moments are for the most part formed by moons. Individual planetary angular moments of the Oss are fixed and they do not all point exactly the same way even though collectively they all average out to point into the galactic arm.

●●●●○○○ Gravitation

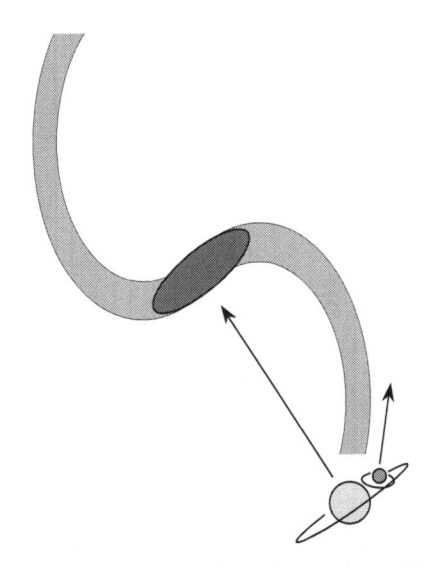

Direction of primary solar and secondary planetary angular momentum

The galactic arm is many solar systems thick and because each of the planetary angular moments individually points to one neighboring solar system there is much spread associated with all secondary angular moments. In the bar of the bar galaxy the spread could be much tighter and the primary and secondary moments overlap overall. Since the moons carry the most of secondary angular momentum by far, the direction of the lunar angular momentum is the best indicator of another solar system that is gravitationally connected with a particular planet.

In a compound two-body system, the lines of force can be anywhere along a two-dimensional plane and this results in force having two dimensions of independence. In a galaxy the lines of force are fixed because a galaxy is a one-body system. However, radially symmetrical movement of solar systems as a whole continues to uphold a single body topology, and this means that every solar system has a "sister" solar system that allows the radial symmetry to hold. In a spiral galaxy the lines of force radiate in one plane but do not move relative to each other. The question to answer is: Are there enough force lines to make it into a dense two-dimensional plane? The answer is no. The force thus claims 1½ degrees of independence. The primary angular momentum of every solar system in a spiral galaxy points toward the center of the galaxy but there exists no mass along this line. Mass accumulates along the path of the secondary angular momentum and thus mass claims ½ degree of independence. This is based on a similar question that asks: Are there enough solar system points to make it into a dense one-dimensional line? The third and last degree is claimed by distance that allows the entire assembly to rotate in any one direction. A spiral galaxy is the hyperstate 1½,½,1.

A spiral galaxy's plane cannot altogether fill in with mass because at that the mass and force and distance would exceed three degrees of independence and the galaxy could not remain computable. Matter formed inside the spiral galaxy will be created in the arms or will move to align itself within the arms. The arms of the spiral galaxy will not attain the star density of a bar in a bar galaxy, for solar systems in a bar of a bar galaxy take one full degree of independence. A good guess is that the more spiraling arms a galaxy has, the fewer solar systems such arms would have.

Ring and Rings

It is technically possible for all of the parameters' degrees of independence to add up to less than three while producing a computable context. However, the competition among the three variables allows any of the three to have as many dimensions as computationally possible and thus all three available degrees of independence are always taken. This does not happen at the atomic scale because, as will be discussed further on, the computable states can exist concurrently and in superposition, which is something the exclusive-acting matter could not physically implement.

When objects are small and do not exert gravitational force at each other – even though they are next to each other – then force cannot claim a degree of independence and exists with zero degrees of independence. If such objects line up in a straight line (taking 1D of distance) and are not subject to external forces then such string of matter is computable in hyperstate 0,1,2. A ring of small objects in geostationary orbit would be a manifestation of this hyperstate and, in addition, the ring should be able to move in two degrees of freedom while remaining computable. A single thin 1D ring, then, should be able to pan and tilt along or around the large body it surrounds. However, since the force-free state is a result of the geostationary orbit afforded by equal body and ring rotation, the entire ring by itself will not move outside of the equatorial plane because it would become subject to forces and would then need to fit another hyperstate. If for any reason the planet were to tilt and pan, or continue to wobble, the thin 1D ring will move and remain computable in the equatorial plane. In the absence of the planet's tilting or panning movement, the mass can increase its claims to two degrees of independence by filling in mass in a 2D plane where the ring becomes a disk. Although there is initially only 1D available to mass because there is only one altitude available for force-free geostationary orbit, mass can accumulate and grow higher than the geostationary orbit because each thin ring of matter adds mass that moves the geostationary orbit higher and higher while keeping the net force on each piece of ring at zero or close to zero. Disk or discus topology is the double-ringed hyperstate 0,2,1. A tilt in the axis of the planetary rotation will computably change the non-rotating (geostationary) disk to align it with the new planetary tilt but a combined tilt and pan would again reduce the ring's thickness to 1D – the disk would break up and the hyperstate changed from 0,2,1 to 0,1,2. If distance were not left with one degree of independence then a tilting movement would result in the entire disk's breakup.

A single 1D ring – that is, a thin string of matter, can exist as hyperstate 0,1,2 around a planet or sun that experiences significant two-dimensional wobbly movement in its axis of rotation. In the presence of simultaneous pan and tilt movements of the planet or sun, the distance parameter takes two degrees of independence and only a radially thin non-rotating 1D ring of mass can make a computable hyperstate. In reverse, it can be surmised that planets with radially extensive rings are stable as far as their spin is concerned.

Quantum Pythagoreans

Summary

Jack spins the rim
Spin, Jack, spin
Watch the wheel go
  riding the virtual wind
For the wind must go
To fill up a hole
That was a hill before

Mapping of hyperstates leads to the working principle of the organized universe: At least one hyperstate must exist within any observational boundary. Hyperstates can be compounded as when the boundary of Oss is increased from Mercury on out, for example. Hyperstates can also overlap. When approaching a galaxy there will be a region where one hyperstate diminishes while another one arises although such transitions may be rapid. A hyperstate can also be contained within another hyperstate: Several planetary orbits together form one or two hyperstates but, taken separately, they each form another, third hyperstate.

If computability does not hold, a chaotic subsystem with potential for collisions results on the cosmic scale while on the atomic scale the effects of ionization, core breakup, or matter-antimatter creation can be expected.

Another way of looking at real structures is by classifying all topologies as one-body or two-body systems. One-body topologies are interesting because these manifest as a gaseous body of a sun or a gas planet, solid body of a planet or moon, separated but distance-locked bodies of a flat galaxy rotating as one body, separated but orbit-interlocked plurality of planets or moons that can be computationally reduced to one body, and separated but symmetrically radially-oscillating bodies of a spherical galaxy where the radial distance is increasing and decreasing as one body. If you were the Creator you would ask, "How many ways can I make a one-body or a two-body system?"

"Today," you might say, "I'll make a one-body system of a hollow sphere of a giant red sun."

It is likely that the gaseous one-body system has inherent friction mechanics where matter accumulates uniformly and gradually, firing up as a sun at some point. It is also likely that a solar system evolves from a single sun that forms heavier matter in quantity and density that is influenced by neighboring solar systems. For example, a production of neutron-rich helium could increase the creation of spin of the solar system. A decrease in acceleration of a solar system calls for an increase in rotation and the creation of angular momentum. The early sun is mostly gas and the rotation then splits the sun into two suns of about the same size. If the sun contains heavier elements, planets are spun off that preferentially consist of only solid material that has inherently greater angular momentum in closer orbits. If a particular additional angular momentum is needed, there may be but a few computable orbits available and then just a particular amount of mass would be spun off that results in the desired angular momentum. It is also conceivable that several planetary orbits may be rearranged if a simple solution is not available. Planetary orbital interlock, for example, allows planets in tighter orbits. Secondary angular momentum is needed when a neighboring solar system is exerting increasing force. A new body or an existing planet is inserted into a larger planet's orbit thus becoming a moon that mitigates the increasing force from a neighboring solar system. From time to time planets or moons can become dislodged, appearing for years to dance the game of musical chairs before reaching another, more appropriate orbit. The creation of a new planet is a spectacular event where a fireball tangentially and slowly rises from the sun's equatorial surface, moving and spinning in ever-increasing orbit as its fiery umbilical cord slowly withers away.

The one-body topology of a galaxy is computable for many specific solutions but the objective of the one-body arrangement in and of itself does not reveal the proportioning of force, mass, and distance among the three degrees of independence that are fundamental to achieving computability. A galaxy contains many different hyperstates depending on the observational size while the galaxy, as a whole, has a one-body topology. Looking at one-body structures as analogies, the familiar solid, liquid, and gas states of matter appear to have some correspondence, although the liquid state of an oscillating or "breathing" spherical galaxy is a bit tenuous. Flat galaxies stand on their own because these are rigid along the plane yet they are flexible to bend. Working the flat galaxy analogy in reverse, perhaps another state of matter in addition to solid, liquid, and gas is a strong yet flexible sheet or fabric of matter.

The nonlocal computational aspect of quantum mechanical gravitation yields scale independence. Scale independence then makes spatial distances and the size of the organized universe unbounded. An identical hyperstate exists for a moon's orbit as for a planetary orbit, dual suns, and two flat galaxies that counter-rotate along one axis while orbiting each other. A small body of a spaceship moving in empty space is in the same hyperstate as a galactic supercluster that is moving in a straight line because there are no other superclusters in the neighborhood.

Spherical galaxies are "loners" separated from other galaxies to a much greater extent than flat galaxies that possess large angular momentum. It is apparent that a spherical galaxy solution can happen when the angular momentum is not needed, which happens in the absence of linear movement. A galactic system with nearly spherical shape does not have a fixed or large angular momentum. A flat galaxy, however, is subject to linear motion that, in consequence, results in angular momentum because the angular momentum arises in the direction of motion and the galaxy's flattening maximizes the galaxy's angular inertia.

Alchemy

Alchemy is about any and all and usually weak associations. The strangeness of relationships may lead some to believe such relationships are irrelevant. The basic premise, however, is that individual influences may be weak and may not even drive the outcome, but together a large number of influences may add up to a decisive force. The influences, moreover, are not qualitatively identical and the final result is not a simple arithmetic sum of all influences. In alchemy, the supporting breadth and the variability of influences are sought and appreciated. Moreover, the influences that are identical for the sake of agreement are mere copies and that of a follower – rather than the actual influences that come from a leader.

Further still, the influences are recognized to operate and are leveraged by particular symbols and geometric structures. During Galileo's time, the Catholic Church taught and enforced the standards of propriety and it was not wise to speak openly on alchemy. Kepler, and to a great extent Newton, used alchemy to form a personal hypothesis, which would then be validated by observation or by experiment prior to being published. The moon's orbit frequency being in synchrony with oceanic tides was the main reason for Kepler's gravitational linkage, while the "moist" moon may have been the first hunch that started Kepler on a path seeking additional supporting evidence. Newton, it is said, made a correspondence between a falling apple and a "falling" moon. It would then be easy to belittle Newton for thinking that "the moon is falling." Since the moon is 60 times farther away from the center of the earth compared to the apple, Newton validated his gravitation equations by taking the gravitational force exerted at the moon to be 3,600 times weaker than the force acting on the apple. Galileo could also be labeled as the proponent of "flat earth" because his derivation of the parabolic trajectory presumes that the projectile is subject to constant gravitational force, which, in turn, is valid only on an infinitely flat plane. Galileo obtained the parabolic trajectory mathematically where the breakthrough was adding and thus superposing the object's horizontal and vertical velocities. Presently, parabolic geometry is an excellent approximation for suborbital trajectory. Yet the greatest of Galileo's attributes stems from his direct challenge to Aristotelian 'Nature abhors a vacuum' credo. When Galileo postulated that movement could happen in the absence of friction, the vacuum context then enabled the parabolic solution of the cannonball trajectory, as well as the qualitative formulation of the conservation of momentum. There appear to be remnants of Aristotle's "no zero policy" going on today when universities espouse laser's radiation pressure impinging on a mirror surface with equations – but have nothing to offer by way of confirmation with direct experiment and measurement. Something tangible, even if it is made up, is put forth by academia to uphold the status quo, for the value of zero raises many questions and leads to most unexpected breakthroughs. Not unlike infinity, zero intrigues the mind and that, for better or for worse, is not about status quo.

All hyperstates map onto a logical plane. While some alchemical and philosophical literature refers to 'physical' or 'real' planes, discussions on planes are oftentimes about higher, lower, inferior, or similar differentiating planes. It is likely there exist planes that are differentiated by frequency content and a differentiation by octaves then becomes likely. There could also be differentiations based on the incommensurable (irrational) numbers, for these numbers create unique groupings of infinite frequencies. Similarly, an "organic" layer that is common to all living things on earth may have its group of frequencies that, while infinite, has a unique profile in its magnitude representation. The hyperstates triangle, however, is a *logical* plane – a template – and hyperstates are about all possible – that is computable – ways of realizing matter on a cosmic scale. Distance alone, force alone, or mass alone does not determine the computable state. Any of the two variables alone do not result in a computable state. Although the dominance of particular parameters can be identified in each and every hyperstate, dominance by itself is not paramount because any parameter can dominate and be in accord with the Maxim of the real domain. Hyperstates template as a whole supports no particular dominance and hyperstates support all possible dominance of force, mass and distance. The competition among parameters must result in a stable system because the objective in the physical universe is the creation of a tractable system that can then become the foundation for future growth.

The three or more body – at times called the N-body – problem is in general intractable. Concurrent computational abilities of virtual elements altogether seek a hyperstate solution to enable more tractable matter to appear in the universe. Tractable solutions offered from the hyperstates plane results in the proportioning of dominance between force, mass, and distance. Real and virtual entities are inextricably yet separately interlinked in the production of the organized universe. The real-virtual duality also does not have hierarchical relationship because the real and the virtual domains are both qualitatively different. Hierarchical relationship is the property of the real domain that does not consider virtual elements. Hierarchical relationships are fixed and unchanging in the real domain because the real domain that does not consider virtual elements. Dominance or independence in the virtual domain changes with context and does not change if the context does not change. For cosmic topologies the context is synonymous with the boundary of the system one selects for observation.

The triangle of hyperstates is the summing triangle. Hyperstates 3,0,0 and 0,3,0 and 0,0,3 define the outer edges of the triangle. All of the three variables' degrees of independence that are inside the hyperstates triangle sum up to three, and every point in the triangle's area is technically a computable solution. Degrees of independence can be any real number between near zero and three and this includes integers, near-integers, or rational numbers. All real numbers represent real entities that are finite in size and quantity. All real numbers have finite precision, which is another way of saying that the mantissa and resolution of real numbers is finite. Yet another way of saying it is the way of the ancient Greeks: All real numbers are commensurable. All irrational numbers then must be trimmed in some way as these transform to become real numbers representing real things. Because the virtual entities are wave based and contain frequencies, the realization of a virtual entity – that is the transformation of an irrational number – can happen by constructing a finite mantissa with a finite number of frequency contributors that have a whole wavelength boundary. What this means is that the trimming of the infinite mantissa of an incommensurable number is not arbitrary.

With the exception of the spiral galaxy, all cosmic hyperstates are formed with integer or near-integer values of degrees of independence. All degrees of independence issue from the origin that is the virtual domain and the virtual domain is at times called hyperspace[79] or ether or fire. Entities transitioning from the virtual to the real domain are thus *projected* from the origin and onto any hyperstate. It is easy enough to see that symbolically the eye is placed at the origin and at the apex of the triangular tetrahedron pyramid while the hyperstates are at the base of the pyramid. The tetrahedron pyramid has three steps or markings that count the degrees of independence from the origin. None of these steps are physically transitioned because the projection into the real domain corresponds to the instantaneous reduction of the wavefunction. Your ship materializing in the real domain can do so at any point on the real plane. The shape of the wavefunction determines where on the real plane your ship appears. The wavefunction acquires its shape through optical interactions with real

---

[79] Hyperspace is not a different kind of space and all real and virtual entities exist in space. A real entity entering hyperspace is still in the same space but acquires virtual properties and hence belongs to a different domain. Going into hyperspace is then the same as transforming from the real to the virtual domain. In the virtual domain repeatability is not a given and, because it may be more difficult to navigate there, some also refer to the virtual domain as chaos.

things while inside the virtual domain. Should your ship materialize in 0,3,0 it will be inside the sun or inside another massive body and you may want to pick another hyperstate for your ship's materialization.

In alchemy there are three quantities called the philosophic sulfur, salt, and mercury. Since there exist three interacting variables as well as the three degrees of independence, it is not apparent which of the two triads the philosophic sulfur, salt and mercury refers to. There are also quite a few triads in common use such as the animal, mineral, and vegetable. 'Philosophic' is an adjective signifying that the reader should not interpret sulfur, salt, and mercury literally and should think of them as essences. In such case the sulfur, salt, and mercury are likely related to the force, mass, and distance — not necessarily in the same order. It is conceivable to be able to distil the force, mass, and distance variables from sulfur, salt, and mercury. The most likely situation is that sulfur, salt, and mercury are representative of esoteric dimensions, fusions, and characteristics such as fire-vapor, solid-crystal, and liquid-metal. If you think the interpretation cannot get any more esoteric, another complication states that a little bit of the sulfur, salt, and mercury is to be found in each of the other qualities. The upside is that, in addition to hyperstates, other discoveries can be made with philosophic sulfur, salt, and mercury. Some literature equates sulfur, salt and mercury with body, mind, and spirit. Body, mind, and spirit certainly have a relationship such that all three comprise each human being. Sulfur, salt, and mercury can, therefore, be thought of as interlinked archetypes from which additional triads could arise.

The idea of numbers as constructs of creation is that while numbers exist as standalone entities, numbers can become something that is grouped and intertwined, and becomes alive — that is, some self-sustaining and permanent entity is created from numbers. The composition of numbers that creates a living entity is, in the Pythagorean tradition, the Monad. The root of the word monad is 'one-sum.' Photons superposing in one locality would not make a monad, for such arrangement is not self-sustaining. It also appears that the smallest monad is the triad. A monad can be ripped apart and broken up into individual components, but the idea behind monad is not that creation is something irreducible. Monad's permanent or self-sustaining existence calls for a particular grouping and actualization of numbers, which may owe its perpetuation to another numerical property — the reversible transformation.

● ● ● ● ○ ○ ○   **Gravitation**

There is some disagreement on the meaning of the monad. Some authors describe the monad as number one, others as the "first thing." It appears that the Pythagorean Monad is something whole that is not possible to subdivide other than by technical means. A living animal or an atom is a monad. Unlike an electron, a monad can be broken up by force, but the idea is that the monad is something that is created to be self-sustaining and possibly self-reproducing. The monad is a single functioning entity that can be seen as a unit that is whole. In the Pythagorean tradition the monad is also described as intelligence – because "intelligence" enables the monad to function as a standalone entity.

In the golden days of alchemy the sulfur, salt, and mercury were some of the basic and common[80] ingredients of work, and the recording of processes in obscure ways was the order of the day. The stated purpose of obscuration was to keep the knowledge from people who could not appreciate the fine art of discovery and who may be ready to steal or corrupt the fruits of one's labor – summarily referred to as the profane. Symbolic documentation, then, offers a quick way of recording findings that are easy to decode once a person spends some time in the alchemical tradition. It is also likely that a particular symbolic or cryptic way of recording the processes is necessary because definitive methods take away the effective reach of the processes. The alchemist knows that the benefits are applicable to several situations, some of which may be in the future. For example, the process of projection is the highest form of transformation and may well have been felt and possibly practiced hundreds of years ago – yet the instantaneous reduction of the wavefunction that materializes a virtual entity was not mathematically formulated until recently and the actual application on a macro scale is some years away.

In alchemy, equal proportions of sulfur, salt and mercury is supposed to result in gold and the thought behind it is that once the projection is mastered one could make more gold out of a little gold. The center hyperstate 1,1,1 is formed with full and equal integers, and results in

---

[80] Sulfur, salt and mercury are as common to alchemists as force, distance, and mass are to physicists.

the barred galaxy. There is nothing golden about the barred galaxy, but gold is highly figurative in alchemy because the growth in wealth – for the individual as well as for the society – happens along with the growth in relevant knowledge.[81]

In alchemy the associative symbolism holds center stage. Metal is the soul of man. The transmutation of metals as well as growing metals into more noble metals is then a process of advancing the soul. The increase in knowledge on how to go about "improving metals" can well be compared to knowledge regarding computability, for metals and atoms in general are computable entities. The alchemist of the 17th Century had no knowledge regarding the mathematical definition of computability but the alchemist would have wholeheartedly agreed his work is about new solutions. To "strike while the iron is hot" is foreign to alchemy, for such process is akin to indoctrination. Fundamentally, the alchemist seeks superior ways of accomplishing work, and the rejection of indoctrination is based on knowledge of better ways of imparting and using the art. Alchemists work with associations and they love to ply words to distill all possible meanings from them – all puns intended. A template can become temporal plate or temporal plane or a time plateau or a temple – each having several literal and figurative associations. Yet a particular association carries no proof – only an influence. Further, a discourse on all possible meanings and applications of alchemy shifts the associative process from the infinite domain of the right brain and into the unbounded domain of the left-brain. When describing relationships with all possible modalities and nuances, the exhaustive and sequential processing then replaces concurrent processing. Not all is lost, however, for many a discourse puts the reader to sleep.

Hyperstates computability is also applicable at the atomic core and it is possible that equal proportioning of force, mass and distance – or their equivalents at the core – yields unique materia there. Overall, the alchemical projection results in a computable state having the benefit of repeatability; the application of leverage could account for 'making more gold out of a little gold.' Then again, equal proportions of competing

---

[81] Bar, however, is a strong phallic symbol and the story of Isis putting Osiris together with a fashioned golden phallus makes it a very deep story indeed.

variables results in a hyperstate where no variable dominates. The proportional balance of dominance between the executive, legislative, and judicial branches of the government may yield a "golden" and stable, though competitive, system.

The three variables competing for the three degrees of independence gives rise to a platform for numerology associated with the number three. Pythagoras went as far as to start all numbering with the number three as the smallest practical number that is the monad. The reasoning for skipping numbers one and two, however, could also be made on simpler grounds of spatial dimensions, for three dimensions are needed before matter can exist. Numerologically, number one is common to both the triangular and square numbers and forms the starting point or seed for other numbers. The number **1** of the Pythagoreans is an incrementing entity. Number **1**, for example, could be compared to the Planck constant, which is the smallest unit of the energy-time product from which any higher energy-time amount is constructed by continuous addition of such units. A geometric point in space is also one item that serves well for rotation and the symmetry about a single point. If you think that there is more to **1** than a number **1**, then shake hands with a Pythagorean.

All square numbers have a high profile. In physical systems any multiplication – that is, leveraging and squaring – gives rise to force and thus multiplication and squaring is related to force. If the force results in motion then multiplication is also about energy. In previous chapters, multiplication was identified as an operation that facilitates transformation between real and virtual domains, particularly since matrix multiplication offers combinatorial interplay between all values a variable can have. Indeed, the transformation via matrix multiplication emulates the workings of the atom where forces arise in response to photonic absorption, for example. In the Pythagorean context the square root of any square number is an exact and a *real* number. Pythagoreans then likely thought of all square numbers as being the "portholes" or "points of contact" into the real domain. Moreover, the diagonal of all geometric squares carries an additional and important property of squares. A square area of size **9** has a side of **3**, but its diagonal is **3√2**. Any square of side **a** has a diagonal of **a√2** and because every square has a constructible 45 degree angle, *all* square diagonals may overlay a single centerline. The root of two and the root of five are prominent in the Grand Gallery of the Great Pyramid – see the Appendix.

Number thirty-three should be seen as two threes rather than the count of thirty-three. The interplay of three variables (force, mass, distance) and four integer values (zero to three) of three degrees of independence results in ten hyperstates.[82] Thirty-three, as related to the number of mammalian vertebrae, is likely related to the vibration nodes of the spinal harmonics, which also account for charkas; spinal energy centers.

The sum of all dots in Tetractys adds up to ten because there are ten major hyperstates or ten major topologies. All virtual entities superpose through addition yet superposition is not a permanent state in the virtual domain unless and until the virtual entity forms a tractable closure of an embodiment. For Pythagoras the number ten was 'everything,' as the sum of all dots in the Tetractys yields ten. The 'everything' of the Pythagorean Tetractys is about all tractable, and therefore formal, systems. Virtual entities find closure in the topologies of hyperstates and can then support the creation of formal systems. The triangular number ten is symbolic of the finite yet unbounded topological solutions that in general make up the reality of the universe from integers, near-integers, and one fraction.

The number three has a very high association with real, physical, and tangible things. Inside the atom on the micro scale the degrees of freedom cannot exceed three if the atom is to remain computable. On the macro scale the degrees of freedom also cannot exceed three if chaos is to be worked. Newton's pursuit of explaining gravitation vacillated on the inclusion of ether in the gravitational mechanism. Ether, as an all-pervasive sea of uncommitted electrons, does not figure in gravitation directly but provides the uniform medium, which, in turn, guarantees the absolute speed of light and the absolute spatial distance, the absolute gravitational force, and the conservation of energy. Ether, as an all-available electrons resource, supports the computability and the existence of real matter.

The number four has a high association with the virtual domain, for the virtual domain is about relationships and at least two variables are needed to begin any relationship. Cartesian coordinates are the

---

[82] The combinatorial interplay will be expanded later as Projection.

mathematical representation of the crossing of two lines. Cardinal directions or a crossroads result in the same framework. The relationship between two double-ended variables happens in four quadrants, as the zero value of both variables is placed at the origin. Virtual variables are always double-ended. The crossing of two lines symbolizes a relationship and, for example, a symbol of two crossed swords does not portray a contentious relationship. Additional and potentially an infinite number of variables can be placed on the coordinates and the relationships among any and all variables can then exist in superposition. The rose or the lotus flower with a large number of petals symbolically represents the infinite superposition of intangible elements. The rose is placed at the heart while the lotus flower is placed at the right brain. In the upcoming chapters, the Cartesian coordinates are folded from a plain to form edges of the four-sided pyramid, as we pursue computability in more depth.

Alchemists also work with numbers and Western alchemy in particular addresses the mystery of three versus four. Not only each of these two numbers represents the fundamental component of nature's duality, but the interpretation of three and four can be confusing as well. The idea is to understand the reality as existing in 'three dimensions,' rather than applying the 'four points' that provide a boundary for volume. The basic and common mistake is to settle on number four as being "material," and then go with number three as "spiritual" because there is no better number around. Not so. The number three is the triad that is a very real thing while the number four deals with the infinity of variables in the virtual domain. If the degrees of freedom are limited at three then real and repeatable (formal) systems can be built. With one or more additional degrees of freedom, we enter the virtual domain, which consists only of virtual entities. In the virtual domain it is not relevant to ask 'how many degrees of freedom there are,' because all virtual entities exist in superposition and relate to each other through their own ranges of dependence and independence. It is most difficult, but not impossible, to find bearings in the virtual domain, for anything is possible in there.

Numerology is a form of symbolism that need not be confined to integers. Irrational numbers can also become real but, since the irrational number contains an infinite amount of values in an infinite number of decimal places, the irrational number cannot transform the entire amount of its virtual energy into real energy. A real entity is always finite and then "a little tiny bit" of the virtual, high frequency energy remains behind because the total energy is always conserved. Ouroboros is the prominent symbol here. By chopping off a bit of its tail, Ouroboros shows the irrational number associated with a circle (spherical or curving geometry in general) that is transforming into the real domain while leaving some energy behind. In reverse, real energy can transform into virtual energy under rotational geometry but some of the energy needs to be added from within the virtual domain to satisfy the conservation of energy.

Circle

Perhaps the most misunderstood symbol is also the one that is quite common: the circle. The circle is referred to as the 'whole' and the 'infinite.' The circle is also seen as zero but as zero the circle is empty and full of nothing. Others see perfection in the circle; by association with the sun and the moon, a circle acquires fundamental qualities of an archetype. The circle is called mysterious and divine, a glowing orb in the UFO lore and in the Bible. A circle provides closure and the repetition of the seasons, but there is no need to be stuck in the circle going round and round without getting anywhere. The wheel of fortune associates rotation with the ups and downs and also as taking the good with the bad. The periphery of the circle can be unwound into a straight line but such length is not an exact number because the circumference of the circle includes the transcendental and incommensurable number π. Yet, the incommensurable distance of a circle's circumference is then well suited for waves of the atomic orbitals, for many different wavelengths can fit between two nodes of the circle, while each wave can start anywhere along the periphery while ending at the same spot on the periphery. But the biggest surprise comes when examining the circle from the perspective of symmetry. The circle is symmetrical about a point but the circle is also symmetrical about a line. The circle is *not* differentiated by symmetry about a center point or a line.

●●●●○○○ **Gravitation**

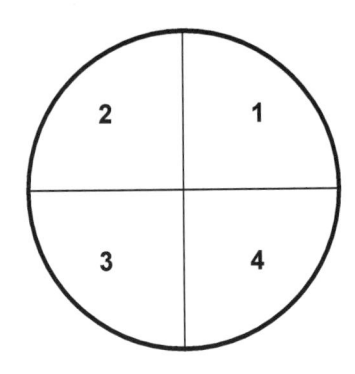

                2 | 1
                3 | 4

    The circle is an even function for the upper or lower half of the circle but not for both. The horizontal axis should have some unique properties because it delineates and transitions between the upper and the lower half. Yet the circle is an odd function, for in diagonal quadrants of the circle – that is, in quadrants 1 and 3 *or* 2 and 4 – the circle segments are symmetrical about the center point. Even symmetry about the vertical axis is a symmetry that is thought of as feminine and inclusive. The odd symmetry about a point is thought of as masculine and exclusive. The circle, then, is somewhat of an enigma instead of being a specific and finished thing. The circle is closer to being a form or a template. A circle is not something that is both masculine and feminine but, instead, a circle is not disagreeable to either and becomes a mediator between the two. Because the particular quadrant segments are acceptable as both the even and the odd function, the circle is also the fundamental initiator and differentiator of gender partitioning. For example, and taking quadrant 1, the only way to extend this quadrant as masculine is to add quadrant 3. But to extend quadrant 1 as feminine, then only quadrant 2 can be added. In our example, quadrant 1 can be shared but the other quadrants differentiate. To an alchemist, the circle can begin to differentiate into masculine and feminine as illustrated below.

Quantum Pythagoreans

Masculine   Feminine

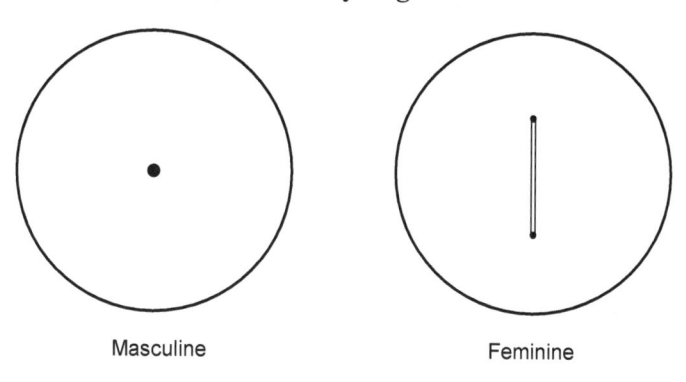

First differentiation of a circle

The masculine differentiator is a solid dot that becomes the zero-dimensional point of symmetry. The feminine differentiator consists of an empty slit while *inside* the slit is the vertical axis of symmetry. The feminine axis of symmetry is the virtual line.

Functionally, a circle is the *delimiter of context*. A circle (or ellipse or egg) defines the boundary and thus the content of the circle. The circle may be evacuated and empty to signify zero or the circle could be a selective barrier for admitting desirable entities while excluding undesirable entities.

Intellectual Property

Pythagoras started a school in southern Italy. He and his graduates were certainly thrilled and delighted when uncovering the fundamental relationships of nature. In the absence of copyright and other intellectual property concepts – as well as the enforcement to go with it for the benefit of all inventors – Pythagoreans practiced secrecy and oaths of silence on the subject of their findings. Mathematical revelations were gifts from God that belonged only to their community. No one individual had the right to release family jewels outside the community. While Pythagoras was the leader to the extent he and the school were referred to as HE or HIM, Pythagoras did not claim ownership personally and no written record authored by Pythagoras is known to exist. While the admission standards were high and biased toward the Pythagorean view of the world, women were not excluded from attending. Pythagoras' teachings, school organization, and his likeness were captured and perpetuated. It can be said that Pythagorean teaching influenced the generations to the extent that the very definition of Western civilization stems from the Pythagorean tradition in mathematics and music.

The evolution of the intellectual property law is unique to the West. Before Newton designed and built his reflecting telescope, other reflecting telescopes were proposed that would use a small curved mirror as the secondary reflector. However, nobody did nor could make such a telescope because the manufacturing tolerances were beyond the capabilities of the times. A device that could not be built and could not, therefore, be realized cannot preempt Newton's actual device, even under the present day patent law. It is conceivable that someone could advance the technology and a patent filing could be made with the expectation that someone would resolve the issues by the time the patent is granted. Yet the presumption of a working model exists at the time of the filing and the working model demonstration can be enforced by the patent office. In the end, while the curved secondary mirror preceded Newton's design, the subsequent incorporation of the curved mirror could make only a narrower claim and only for the improvements over Newton's realized design. In addition, improvements cannot be patented if they are revealed and enter the public domain. Further still, an improvement must also be an invention, which requires that the improvement is not obvious to persons who are familiar with telescope design and construction. In the West, only the legal owner of the invention could apply for a patent.

Presently, there exists a challenge to the patent law. As intended, inventions that affect life itself cannot be patented; for example, food cannot be patented. Yet the occurrence of genetically modified animal feed raises the issue of patentability if the genetically modified feed becomes, directly or indirectly, a component of human food.

A mathematical equation by itself cannot be patented but the author could obtain copyrights for his or her work. Guillaume de L'Hospital hired Johann Bernoulli on a part-time basis to mathematically address issues L'Hospital felt were important. L'Hospital published his own and Bernoulli's mathematical work when he felt that a particular mathematical result was new and beneficial. L'Hospital gave credit to Bernoulli but did not seek his permission to publish. One can make a case here that the authorship and ownership of intellectual work became differentiated with this precedent and to this day the mathematical test Johann Bernoulli worked on is known as the L'Hospital Rule.

Pythagorean lore includes a story of punishment for the disclosure of internal knowledge. If one believes Pythagoreans were closer to a mystical cult than to a competitive research organization, it is unlikely Pythagoreans would resort to the mundane methods the likes of drowning. It is more likely the former member would be ejected and then symbolically buried, possibly with much hand waving and proclamations that oath breaking carries a penalty. Presently, an unauthorized disclosure of information can draw as serious a charge as treason, and it is for this reason that the Pythagorean lore of punishing someone with extreme prejudice will not fade away. Pythagorean secrecy makes for a good story, too. Pythagoreans would likely say nothing to dispute some notions because their 'Truth Prevails' core belief should in the end carry the day.

It is also up to the reader to do some detective work, particularly in the presence of conflicting information. Several sources describe Pythagoreans as vegetarians, for example, which was unusual for the times. The report that Pythagoras sacrificed oxen upon his discovery of the triangle theorem then carries little weight and in fact places doubt on other statements made by the same source. Other reports may attempt to diminish Pythagoras' discovery of the right triangle theorem by pointing out that Babylonians many hundreds of years before Pythagoras described triangles called triplets such as the triangle with sides 3, 4, and 5. What makes the Pythagorean theorem a great discovery is that it works for all sizes of right-angled triangles rather than just for sides of some particular lengths. By generalizing the relationship among the triangle's sides, one can make a case that Pythagoreans enabled the rise of the science of arithmetic and the use of letters as representing any length. The fact that the Pythagorean theorem embraces both rational and irrational numbers is most advantageous.

Pythagoreans were much advanced for their times and were thus a source of suspicion. The unruly mob figures prominently in Pythagoras' demise, either as a precursor or a back-fill to the Christian and Muslim mobs burning down the library of Alexandria. The Byzantine and Roman emperor Justinian will also be remembered as the head of state who, in 529, shut down the longest running institution of learning in the world: Plato's Academy. Yet, the challenge to Pythagoreans and to their advancement is no less than to other groups whose members are dedicated to breakthroughs – be it in physics or magic.

Core

The atomic core consists of but two components also called nucleons: protons and neutrons. Neutrons and protons have nearly the same mass. A proton has equal and opposite charge of the electron but the proton is almost 2,000 times heavier. A neutron is a spinning proton-electron pair, which explains its neutral charge and its weight, which is somewhat larger than a proton's. Placing the neutron outside of the core, in about fifteen minutes the neutron breaks out into a proton and an electron. It is apparent that a neutron maintains its cohesive spin only in the computational environment of the core. An electron may pass through the core with near-zero amplitude of its wavefunction because each electron maintains radial symmetry about the center of the atom while maintaining one contiguous spatial volume.

The eigenstate of the atomic electron is event-driven and the external stimuli account for the electron materializing in an eigenstate. The eigenstate of the core, however, is periodic and every time the eigenstate comes to turn, the linear and angular momentum manifests at the core. One way to test the role of neutrons is to place neutron-rich material in the gimbals and observe the commencement of spin in the frictionless but gravitational environment of the low orbiting satellite.

Core components periodically transform between real and virtual domains. In each domain the computational modality of nucleons changes to sustain or to reorganize the atom. At each transition into the real domain, a quantum of momentum is realized that manifests as the gravitational force or weight of matter. Protons realize the linear component of the gravitational force while neutrons realize both the linear and the angular component. The frequency of transitions is generally constant because the weight of matter is constant.

Quantum Pythagoreans

Continuum of Energy and Direction

Upon acceleration the inertia engages to conserve the work imparted onto a real object. The application of external force over some distance invokes the storage of the applied energy via the inertia – technically called *work* – which is local to the body and operates in addition to the gravitational mechanism. Inertia mediates the expanded work while the body is being accelerated. The work is saved as the increase in vivibration. Similarly, a gravitationally accelerated body also acquires an increase in vivibration.

Work is applied force over a particular distance. The salient point of work is that the applied force over distance has certain direction and the direction is conserved as well. Work, then, is the *energy-direction continuum* where the energy component guarantees the continuum, and therefore the direction, to hold. If a body is imparted with some work, then such body continues at that particular direction and at another point in the universe the body will perform the same work while conserving the direction as well. The spatial direction is the absolute parameter because all observers, now and later, can agree on the direction the body is heading. At the time the energy is imparted on the body the direction becomes known to all observers and continues to be conserved in absolute terms.

Leibniz first recognized a flaw in Descartes' work and pointed out that during collisions not only energy but also the direction of moving bodies is conserved. This was also understood by Newton who, perhaps independently, addressed the conservation of direction by postulating force as a physical parameter that changes the direction of a moving body. Technically, however, it is not force but *work* that changes a body's direction. Force must be applied over non-zero distance if force is to become work – that is, energy – and a body's path is changed in consequence. On level surface there is gravitational force acting on the object but because no work is expended the object does not change speed or direction. Newton's law, then, should indicate that distance of some direction is also necessary in addition to force if a body is to change its speed or its direction. But even with all technical corrections applied, it is the energy-direction fusion that forms the continuum of energy and direction – and the recognition of such fusion is important in its own right. The way to visualize the fusion is that energy is stored as the wave of

259

● ● ● ● ○ ○ ○   Gravitation

vivibration, which is local to the body because it is imparted on the body but, as a wave, it is inherently nonlocal and the energy that is vivibration is not confined to a point. Wave's nonlocal existence then must also have some spatial direction in which the wave exists as it vibrates. While Leibniz did not consider energy as the necessary ingredient – the crucial glue – in the formation of a continuum, Leibniz was a great proponent of a continuum as a mathematical as well as a physical concept. It is then appropriate to dedicate the postulate of the energy-direction continuum to Leibniz, as follows: **(1)** Commencement of movement of a real mass body forms the energy-direction continuum; **(2)** The energy-direction continuum is conserved; and **(3)** Energy is the necessary component in the creation of any continuum – that is, no continuum can be created without energy.

'Real mass' in the above definition signifies that a body's inertia is the agent that conserves the energy-direction continuum. In the general physical context, the procuring work always includes the parameter of direction and then the energy-direction continuum is always created. If the supplied work is angular (symmetrical about a point) the direction of energy-distance continuum is in the direction of a right handed thumb. Presently, the standard unit of work, or *force* times *distance*, is a *Newton·meter*. The old English unit of work is *horse*. Neither notation, however, conveys the directional aspect of work and deals with but a magnitude. A new English word that captures distance *and* direction is needed. Perhaps *directance* could fill the colloquial void. Asking for directance would make sense because it is nice to know in which direction and how far it is to the Faneuil Hall in Boston without relying on "you cannot miss it."

Projection

On a cosmic scale the competition among the three variables for the three degrees of independence is a process of taking the three variables and *projecting* them with the integer values of degrees of independence. Lets postulate the Projection operator ⋈ where **n** ⋈ **d** is the projection of **n** variables with **d** degrees of independence. For the case 3 ⋈ 3 the projection results in 10 states since the three variables competing or combining for three integer values of independence yield ten (hyper)states. The projection

260

Quantum Pythagoreans

operation can be extended to any number of variables and to any number of degrees of independence. The following table lists the projection of two to four variables, each with zero to three degrees of independence. There is no need to go beyond three when dealing with degrees of independence, for any higher value renders the system intractable. In general, the summing total of degrees of independence can also include non-integer values of degrees of independence. The projection of a single variable is not available due to the first principle where no single variable can attain independence under all circumstances.

Number of Variables [n]	Degrees of Independence [d]	Projection States n ⋈ d	States
2	0	1	0,0
2	1	2	1,0; 0,1
2	2	3	2,0; 0,2; 1,1
2	3	4	3,0; 0,3; 2,1; 1,2
3	0	1	0,0,0
3	1	3	1,0,0; 0,1,0; 0,0,1
3	2	6	2,0,0; 0,2,0; 0,0,2; 1,1,0; 1,0,1; 0,1,1
3	3	10	3,0,0; 0,3,0; 0,0,3; 1,2,0; 0,1,2; 2,0,1; 1,0,2; 2,1,0; 0,2,1; 1,1,1
4	0	1	0,0,0,0
4	1	4	1,0,0,0; 0,1,0,0; 0,0,1,0; 0,0,0,1
4	2	10	2,0,0,0; 0,2,0,0; 0,0,2,0; 0,0,0,2 1,0,1,0; 0,1,0,1; 1,1,0,0; 0,1,1,0; 0,0,1,1; 1,0,0,1
4	3	20	3,0,0,0; 0,3,0,0; 0,0,3,0; 0,0,0,3; 1,2,0,0; 0,1,2,0; 0,0,1,2; 2,0,0,1; 1,0,2,0; 0,1,0,2; 2,0,1,0; 0,2,0,1; 1,0,0,2; 2,1,0,0; 0,2,1,0; 0,0,2,1; 1,1,1,0; 0,1,1,1; 1,0,1,1; 1,1,0,1

The Projection operation is not commutative – that is, n ⋈ d ≠ d ⋈ n. For three variables the visualization is in the form of a

261

● ● ● ● ○ ○ ○ **Gravitation**

triangular pyramid while the number of states increases in a sequence known as triangular numbers.

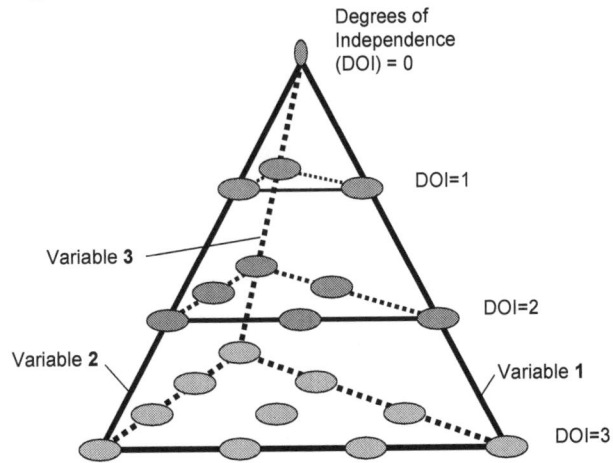

Three real variables projected onto integer values of degrees of independence

For the purpose of projection, the height of the pyramid is logical and measures the degree of independence starting with zero at the apex. The width of the base is also logical and the base is the plane where all possible and individual computable states project. The logical height and width of the pyramid indicates there are no fixed or preferred proportions associated with the pyramid and that the only ratios that are needed are the three steps or gradations of the three levels. The logical rationing relates to computability in general, but specific computational needs will require specific geometries, as is the case in the Great Pyramid.

While the maxim of the real domain limits the degrees of independence due to tractability, there is no apparent maxim on the number of variables. On the cosmic scale, we can link the creation of topological formations to the three variables of force, mass, and distance. The projection of four variables onto three degrees of independence results in a square with twenty states as shown below. Eight inner states overlap into four pairs and all states are at the base of the pyramid where the three degrees of independence are located.

Quantum Pythagoreans

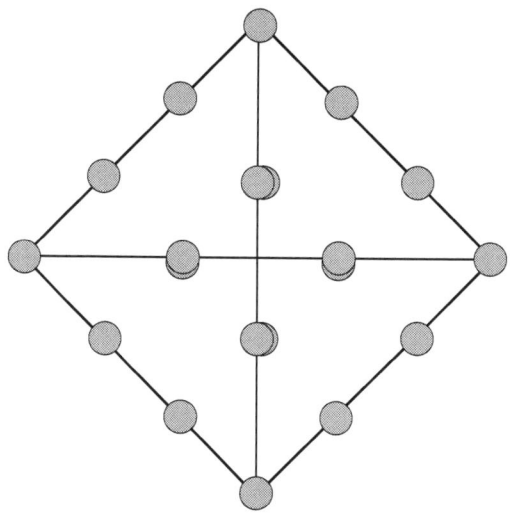

Four variables projected at three degrees of independence

There are twelve unique outer states with generally even distribution around the periphery of the base of the four-sided pyramid.

Real variable can be visualized as a telescoping bar that varies on but one end. Quantity of goods sold moves up or down but is never below zero, for example. When or if a real variable transits to virtual it unfolds with its opposite. While the two opposites are joined together at zero, both ends of this virtual pair can vary independently of each other. Two virtual pairs of double-ended variables, then, comprise four variables. This can become a riddle:

> *Connect two at zero into one*
> *Either arm reaching is two*
> *The Sphinx has a clue*
>   *four is pairs two*
>
> *But the Magician knows*
>   *four on each floor*
>   *one dozen in all*

The answer to the riddle is that while there may exist four interacting variables, two distinct virtual entities operate as four variables because every virtual entity has two changing variables that are joined at one end. This is analogous to having two centered virtual variables interacting in four quadrants of the Cartesian coordinates while overlaying the cardinal axes. Moving this scenario onto the four-sided pyramid, two virtual entities are centered at the apex with their zero value. Each end of the virtual variable originates at the zero-dimensional point of the apex of the four-sided pyramid and extends out along the edges of the pyramid. The values such virtual variables acquire are measured in *degrees of independence*. The four-sided pyramid is about the interaction of two virtual pairs of opposites for each degree of independence thus projecting a total of twelve variables.

Projection of a variable from the apex reaches to either the first, second, or the third level at any one time and represents the degree of independence such variable may have. Moreover, virtual entities can be *branched* and overlaid on top of each other along the edges of the pyramid. The projection of a single virtual variable can then be made against several degrees of independence at the same time. Technically, degrees of independence are not confined to integer values but in the case of hyperstates there appears to be an inclination toward integer or near-integer values. The sum of the degrees of independence cannot exceed three but computability can be had with the sum being less than three. There is a marked absence of competitive bias in the virtual domain, in large part because exclusivity is not a property in the virtual domain. The interacting variables are *not* each striving to acquire all three degrees of independence. Computable solutions among interacting virtual variables are then present at all three levels of the four-sided pyramid and conceivably at any point in the *volume* of the pyramid. The three-sided pyramid can be called the odd pyramid while the four-sided pyramid can be called the even pyramid. In the Pythagorean tradition the odd is "male" and exclusive while the even is "female" and inclusive.

The overlap of the eight inner states at the 3D projection level of the four-sided (even) pyramid is revealing in that the only entities that can overlap each other are virtual entities. A virtual entity is an even wavefunction that allows the infinite superposition to exist without conflict and the overlap represents sharing of pairs of distinct states. This then also leads to the framework of having an infinite number of variables superposed along the edges of the pyramid with the idea of facilitating concurrent computations. The mechanism of reaching a solution would then entail selecting a relevant set of variables from an infinite number of variables that produce a well-defined, well-rounded or unique solution. The challenge is to figure out the computable solution within the pyramid that would then attract particular virtual entities to the edges of the pyramid. It is also likely that a number of particularly spaced pyramids may be needed for more complex solutions.

The outstanding issue is the identification of the actual virtual entities that partake in computational interactions and thus mathematically shape the resulting wavefunction of the virtual entity. In the case of physical matter such variables can be few in number. One may think of light as an example of the virtual – that is even – entity. The photon of light usually propagates and interacts in a straight one-dimensional manner but particular geometric constructs can have light propagating and thus also interacting in 2D or 3D. Light is a candidate for entities that form stable – that is computational – structures. Photons generated by matter, however, do not stop and center easily at the tip of the pyramid but photons of matter can form standing waves or close into orbitals when presented with a suitable structure. It is conceivable a structure of an atom or a molecule can be constructed where suitable photonic energies would be computationally manufactured under pyramid constructs. A photon of light generated by matter does not have its opposite but light can be parted in such a way that each end of the photon can independently interact in up to 3D.[83] A single photon of light, then, can have an infinite number of independently interacting parts produced through a particular geometry. In regard to scale, some ratios such as the Golden Ratio hold regardless of size. It is likely the highly geometric structures that abut each and every pyramid at their east side produce the particular shapes and energies that allow the construction of particular atoms and even more complex creations. Other candidates for virtual entities that can act computationally in up to 3D are the virtual electron that, in turn, produces forces acting in 1D (separated by distance), 2D (spread as discus) or 3D (spread as cloud of some volume). A single electron in the virtual state has two virtual variables – the virtual electron and the virtual positron – forming four variables altogether: just right for the edges of the even pyramid.

---

[83] While this issue is quite complex, a good candidate for the beam splitter is the antechamber of the Great Pyramid.

●●●●○○○ **Gravitation**

When the atomic core is in the virtual domain, the eight inner states and twelve periphery states maintain computability of entities that are free to reach out in three dimensions – that is, the computability of 3D entities happens on the third level of the pyramid. All twenty states exist in superposition during the virtual phase but only one state (only one solution) will become materialized, which reflects the best solution in the presence of external and possibly disturbing influences.

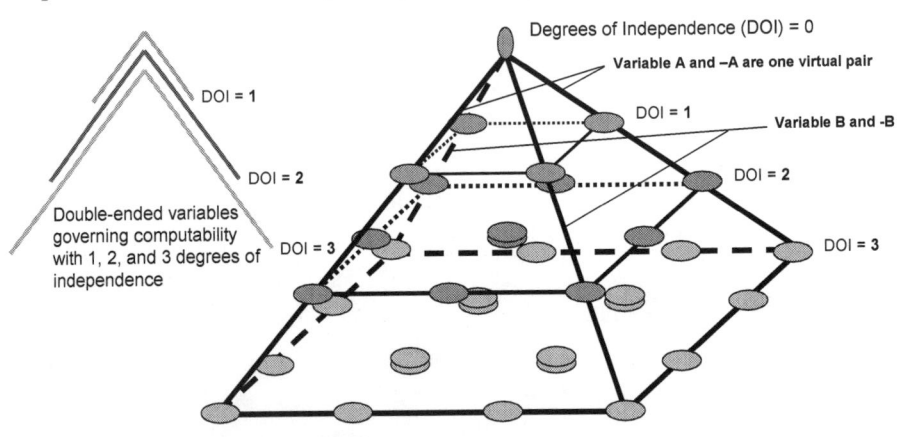

Two virtual double-ended variables projected onto integer values of degrees of independence

Going from *three*-dimensional interactions to *two*-dimensional interactions in the even pyramid is the ascent going up while counting down the dimensions of independence. The full even pyramid is shown above with all computable states of the four-variable projection. There is one instance of two two-dimensional states overlapping each other in the center. There is no strong evidence for this entity but it is likely the two-dimensional force originates at the virtual neutron with a positive-negative (up-down) spin of a dipole that extends its computational reach in but two dimensions of a spinning plane. The eight outer states at the two-dimensional level (DOI = 2) deal with two-dimensional computations, possibly in the gravitational spin context. The result is that neutrons add computational modality to the core that, in addition to or because of the spin, enhances core stability. The torque and consequent spin is computed at this level and materializes during the reduction of the gravitational wavefunction. The gravitational wavefunction contains one of the virtual variables that computes the values for the angular momentum. The magnitude of the angular momentum is also a function of the quantity of the neutrons present. The spin, in turn, allows the creation of structures having two-body topologies.

The first level of the four-sided pyramid appears to be the place for the one, the ancient, and the most influential function: gravitation. Gravitational wavefunction computes and acts in but one dimension that manifests in but one direction of the linear momentum.

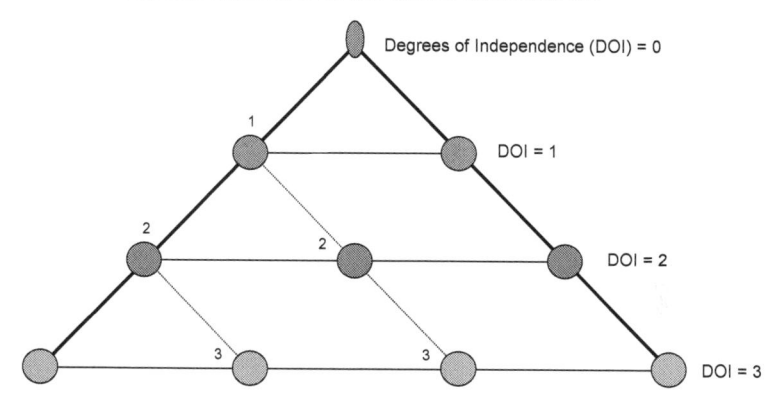

Projection of one pair with integer values of degrees of independence

A free electron is spreading in up to 3D and the virtual electron can be interactively shaped into just about any shape as long as the virtual electron remains in one spatial volume – that is, the electron has fixed energy. Two-dimensional interactions – shown in the "flat" pyramid construct above – would make the electron the shape of a pancake and there are no inherent barriers in having the electron assume a fraction of the degrees of independence that results in a dual hollow cone having, for example, 2½ degrees of independence. Such complex shapes, however, usually happen inside the atom where the electron is subject to complex computational interactions.

On a cosmic scale the variables of force, mass, and distance relate to each other in that either a larger mass or smaller distance, for example, result in greater gravitational force. Inside the atom, nucleons will materialize in any computable hyperstate within the hyperstates template. In the core, the quantized parameter of charge replaces the parameter of mass. Force remains force but is now electrical or magnetic and can be of either polarity. Spatial distance remains distance.

A gravitational eigenstate materializes all nucleons in close proximity and in a particular highly symmetrical configuration. Once real,

267

● ● ● ● ○ ○ ○   **Gravitation**

proton mass along with its charge becomes localized and protons accelerate away from each other. The symmetrical materialization allows all nucleons to accelerate in a tractable, rather than chaotic, manner. As soon as protons start to move they create an electromagnetic field that moves or spins neutrons via neutron's magnetic dipole and neutron tilt likely introduces a spin to the entire atom. Moving nucleons acquire a wavelength and begin to spread as they transit into the virtual domain. Another way of describing the transition into the virtual domain is that the protons' parameter of charge as well as the neutrons' parameter of the magnetic dipole begins to superpose over adjacent volume of space. Parametric superposition allows the nucleons to concurrently and computationally relate to each other. The goal of computational interactions in the virtual domain is to rework the wavefunction of nucleons such that when the next gravitational eigenstate happens the nucleons materialize at the center and with the symmetrical geometry that allows periodic and repeatable pulsing of the core while the core remains tractable. The eigenstate of the core is triggered by the periodic reduction of the gravitational wavefunction. Core eigenstate and the gravitational wavefunction reduction are one and the same, and all virtual attributes become real. This mechanism requires the gravitational eigenstate to be enabled simultaneously across the entire universe. Because the conservation of momentum has priority, in the absence of the guarantee of momentum conservation the reduction of the gravitational wavefunction between any two nucleons does not take place.

ENTRY	270
SECOND SQUARE	273
FIRST SQUARE	276
FROM THE TOP	278
BACK TO THIRD SQUARE	279
GALLERIES	281
INTO THE MIST	282

⁙ The House of Patter and Octavia

My left arm was turning numb from the burning sun as I trimmed the Ohio's lakeshore west to east on the interstate eight oh. I was driving for hours on end, yet the sun was and stayed to my left. There was no need to panic, for it was all part of the excursion – a relaxing discovery drive through the heartland. So even if I were lost it was no big deal, and I could never understand how some people would claim to have missing time – of all things – for the missing time and not worrying about time was my pleasure this day. In the evening, though, the bearings I thought I had gave out. I wish I could say the pilot stayed with me or that the spirit in charge did not check out. I was not running from anything other than enjoying the fast car, you understand, and then, as if the alarm clock went off, the real trooper without much hustle made available to me the institutional lodgings of the state of Ohio with an offer I would not refuse.

Entry

"Is this a real house?" I asked.

"No, no," Mist said, surprised with such a direct question. "The house is on the ground."

"Of course the house is on the ground," I said with some disdain of the obvious.

I came across this composition or a structure that did not look like a regular house. Whatever it was, curiosity got the better of me. Pointy and edgy all around it resembled no house I have ever seen. When I first saw Mist she had a mister in her hand spraying about and that is how I named her Mist. I like to name people and things. Mist kept the place humid and she must have thought it is healthy for people and plants as well. The best thing, it occurred to me now, was not to ask questions unless necessary.

I circled this extraordinary house taking notice. I drew the foundation in my mind and behold, the house Mist talked about was right there.

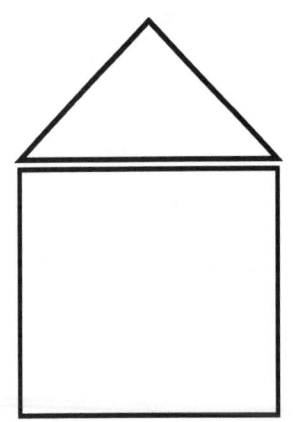

Now this was interesting. Better still, this was mysterious. I liked that Mist could be mysterious. I can be mysterious too, but I also like things straight and to the point. I could hardly wait to talk to her again and ask her questions I knew she would be happy to talk about.

"The house is *in* the ground," I said to her the next time.

"House is the ground," she said.

That floored me. She could easily keep up with me while doing her work no less. Mist was not picky and that was nice. Perhaps I could ask all the questions I wanted. Perhaps I can ask any question.

Quantum Pythagoreans

"Who am I?"

"A child."

Better back off a bit, I thought. I am no child. A child can be amused but I have grown up some. On second thought, amused is better than confused. I will have to put some work into my questions.

I walked around the outside of the house again to collect my thoughts. There were sounds coming in from everywhere – rhythmic for the most part. There was a pacing metronome in there, drums as well. I looked up along the sloping walls of the house and saw a glowing and a flickering flame up at the tip where all walls came together in a point. Fancy place. The gap did not make much sense, though. The triangle and the square were separated not only at the ground but all the way to the top where the flame was. I moved back until I could see the whole house without straining my neck.

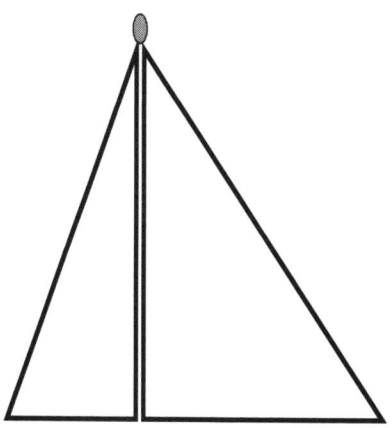

The triangle at the ground was on the left while the square was on the right, both tapering up toward the flame. From this side the house looked like sails of a fast boat. It also looked like a person with a long cape reaching to the ground. It could be a tent as well. The funny thing was that no matter what side of the house I looked at, the house always had a shape of a triangle or several triangles making one larger triangle. Sort of like a letter A. I was now ready with another question.

"Does it go straight to heaven?" I asked.

"Maybe."

Now that was no answer. It either reaches to heaven or it does not. If Mist does not want to be straight with me and give me a yes or no answer then ... then what. Nothing. But I could begin to see how she would answer my next question.

"Are you an angel?"

"Sometimes."

Bingo. I knew she would say that. At times she is, and at times she is not. Or maybe she is an angel to somebody and she is Mist to me. Maybe – this 'maybe' can be a useful word sometimes.

"Do you have friends?" I continued.

"I have many friends."

I knew Mist does not like to give out straight answers, but I still had to ask.

"How many friends do you have?" I said thinking I might get Mist confused for a change.

"Infinity."

I did not like that answer. Did she have a friend named Infinity or did she have so many friends she could not count them all? Or is it that Mist just does not know how to count. I cannot have an infinity of friends and I can count very well. What a sneaky thing to say.

I was beginning to think there was not much I could do with Mist but I was still having fun. There was the flame at the top and there were the sounds and I wondered if things could go across the gap at different heights of the building.

"Who is playing the music?" I said and I was ready for a strange answer.

"Inertia."

"Is she one of your friends?"

"She is everybody's friend."

That was easy. If Inertia is everybody's friend then she is my friend as well.

"Thank you. Where can I find Inertia?"

"She is everywhere so that everybody can hear her."

"How can I talk to her if I cannot see her?" As soon as I said that I knew I did not have to ask that question. I was Inertia's friend and I could hear her.

"Inertia speaks to you when you move faster."

I see. The way I talk to Inertia is by moving faster and she will talk back to me when that happens. I was eager to try it out. I moved toward the stairs and through the humid mist and went up and up faster and faster. I could hear Inertia's sounds getting higher and higher. When I stopped the pitch came back down but the beat did not change. I caught my breath at the next level.

Second Square

It was a hall full of mirrors and it did not surprise me to see ballerina there as well. You could tell she was serious about ballet, for she was spinning pretty in front of her mirrors and I could see her skirt fanning out all around her as straight as can be.

"I just came from downstairs," I said, "after talking with…" And then I stopped because ballerina does not know Mist because I know Mist as Mist and ballerina does not know that.

"You spoke with Octavia. Are you a visitor? Yes, I can tell you are."

"How can you tell?"

"Octavia sends her clients back to the council."

"But how do you know I am not on my way to the council?"

"Because the council is at the triangle and you are in the square."

"Oh." That was all I could muster. As a visitor, though, I can ask some questions and I was fine honing one quickly.

"Is this a public square building?"

"Yes. Everybody can come. Here on the second floor my clients have flat things on their mind."

It did not make much sense to be worried about flat things. I do not iron any more. I was more interested to learn about Octavia than worry about pancakes that were not flat enough.

"Inertia likes to dance like you do," I said, showing my quick thinking.

"She keeps everyone in balance. When Inertia's music gets higher pitch then things get really moving around here. Except on the first floor. First floor does not dance to Inertia's music."

●●●●●○○ **The House of Patter and Octavia**

This is continuing to be interesting, I thought. It looks like many things are happening altogether but there are always exceptions. I still did not understand why they have different floors.

"Octavia is the only one that does not listen to Inertia?"

"You *are* a visitor! The first floor is on top. It is the first floor below the flame."

Finally it was coming to me. It was coming at me like a thunderstorm except that I was not concerned because it was a regular brainstorm. A strange and exciting feeling was coming over my body that seemed to align itself in a new way. I could have figured it out much earlier if only I worked the count. Of course, how could Octavia be named Octavia if she were not listening to Inertia's music? But yes, the third floor is the ground and now that I thought of it the third floor is as far as it goes. Now that I thought of it some more the third floor is where the ground starts.

"I would like to visit the first floor. But I did not ask your name."

"My name is Isotopia. Would you like to stay for a snack?"

I thought over flat food I like because I was certain Isotopia could do no other.

"Pancakes would be great," I said, "…and here is a coin for your hospitality."

"You are a super guy. Come again."

I knew why she would like the coin and why I would come back. She was quite a gal and so friendly. Now I wondered if the first floor person was going to be as nice as Octavia and Isotopia. And I also guessed that gals were in the square part of the house.

The pancakes were delightful and as I expected they were thin, flat and square – with two dabs of butter at the center and more dabs on the outside that looked like this.

Quantum Pythagoreans

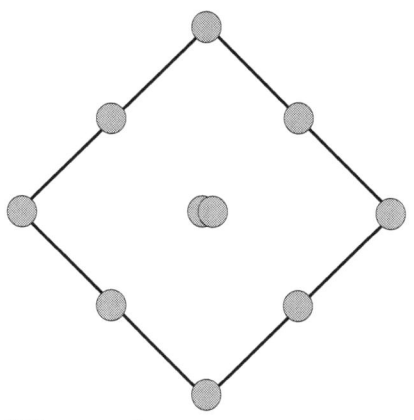

"The pancakes were great. So thin and really flat, too."

"Grandma makes them."

"Terrific," I said, even though I did not see anybody as I looked around.

"We call her G. She makes the platform for the council here because..."

"I know – it's flat."

"Not only that, G knows which way North is and she keeps everyone in the know. G is usually on the first floor taking the latest measurement."

So far so good, I thought. But there was still that gap.

"There is a gap between the square and the triangle buildings and it seems I cannot get across," I said.

"Ready or not, G gives the signal for clients to jump across the gap and into the council seats."

I never would have thought of Grandma making tables, figuring out the North and then telling everyone to jump. But I could see why Inertia would tell everyone how fast they were going.

"You are great," I said to Isotopia. "Do you mind if I stretch here? I would like to rest before going up to the first floor," I added, knowing that at least the bed here would be, well, flat.

Isotopia pulled out a small box and out came a silver bundle. The bundle kept unfolding again and again until it became a thin and a very straight blanket that was as flat as a table.

"Nobody can see us under the blanket," she winked. "Nobody can tear it or cut it but when you are refreshed you just fold the blanket back in the box."

I will have to dream about this one, I thought. Isotopia knows how to make the most unusual things and have fun, too.

First Square

Ascending the stairs I carefully poked my head up in the first floor and was not disappointed. Rays of laser light were bouncing in every which direction. Tubes of light were crossing each other as they rushed their way. All kinds of colors made it a great show. I tried to think of a name for a gal on this floor but could not think of any.

"Is anybody here?" I asked politely.

"Hi, I'm Lux," she said. "My screen name is Twiggy. Do you like my show?"

"Very much," I said wondering why I did not have to ask her name, for just thinking about it had almost the same effect. "Quite different show than on the other floors," I continued.

"Octavia is a witch and Isotopia I would not want to say."

Ha. These gals do not get along very well and that is why they are on different floors. Yet they all do such great things. Twiggy seemed very independent.

"Inertia does not bother you, I hear."

"I do what needs to get done and when it needs to get done and I don't care what Inertia plays. Excuse me."

She was gone. In a flicker she disappeared. Then she was back. She looked a bit different but I had no reason to think it would be someone else.

"Where did you go?" I said, not being sure.

"I run into guys and then I have to split, you know."

I knew she did not have to get my permission to leave. But was she quick. Twiggy was gone and back here just like that.

"I thought you ran into G," I said knowingly.

Quantum Pythagoreans

"We get along just swimmingly and we never bump into each other. G can pull strings, you know."

It was beginning to look G was keeping this place organized like clockwork, ready or not.

"Do you put on shows for guys?" It seemed Twiggy liked to do that.

"Mostly for those guys who like straight things. Like reflection shows. Sometimes we shoot out of the cannon; and the arrows, of course. Most of the time we are cheering in the galleries all around the council. We get in position when everybody gets all charged up ready to fly."

From council assembly to a circus in a blink of an eye, I thought.

"Isotopia dances with guys going around in circles," I said.

"It sounds like a circus but there is a lot of symmetrical formations, precision flying, and synchronized spinning. Isotopia teaches spinning so they do not wobble."

"Wobble?"

"If Isotopia's guys wobble and fall down it's all over. Octavia and I get to pick up the pieces. Usually, my guys get me started and then other guys catch me and we never fall down. There is no up or down once you get good at it."

"What are Octavia's guys doing?" I asked, having no idea what other kind of dance could there be.

"Octavia's guys are the strong guys. They climb on each other's shoulders and come up with the strongest pattern at the council. After the council they fly off to Octavia and she fixes their bruises and sometimes they try to figure out if there are other ways of making different or stronger council assemblies."

I work with patterns. While some structures are stronger than others it is not easy to see which are the strongest. Computers help me but if I get a pattern out of the computer, I do not know if there is another, stronger pattern. So the computer can go on and on but I do not know if there is or isn't another pattern that is a little or even a lot stronger.

"You think Octavia is a witch because she can figure out the strongest patterns," I said with authority.

"She has the place full of mist saying it heals their bruised shoulders, but I think she gets them in a different mood or she is trying to cover up something. Excuse me."

I was still thinking about Octavia and her secret potion when Twiggy reappeared.

"Hi, I'm Lux. My screen name is Twiggy. Do you like my show?"

I was not upset there were more gals named Twiggy. The next Twiggy was just as helpful as the last one.

"Do you mind if I look around? There is a flame I would like to explore."

"I'll do some stretching exercises in front of my special mirror, if you like to watch. Try to catch me if you can!" Twiggy said, adding, "Exploring the flame can take a long time," and she said it as if she was saying goodbye.

"Thanks." I knew Twiggy could go in many directions but she would always come back to one place to finish or to start a show. I knew that once Twiggy started the show she could not rest until she was finished. As fast as she was, when Twiggy said that something takes a long time she really meant eternity. I liked Twiggy and I was not ready for eternity.

From The Top

I peeked my head out from the top and looked around. I drew the structure as I saw it from the top. It was no different from the way you could figure it out if you walk around the foundation and take a look at the sides from the ground.

Quantum Pythagoreans

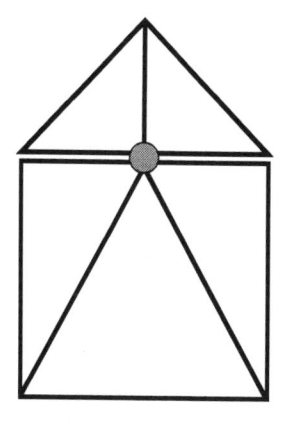

I was thinking about all possible things this view of the house reminded me of. The boat was still there, perhaps a barge or an ark, but it looked more like a keel than a sail. It also looked like a person doing exercises. The flame appeared as a source of some energy or knowledge or force or vitality – something so abstract I could not put my finger on it. I looked into the flame and I knew I could go in but I had to know more before I go in for a visit and come back in time for dinner. Knowing more also included figuring out how to get across the gap and into the council. It was time to see witch Octavia again.

Back to Third Square

Octavia was having a deep-dish pizza.

"Hello, Octavia," I said and waved. "This is a great looking pizza," I continued as I extended both of my arms at the square pizza.

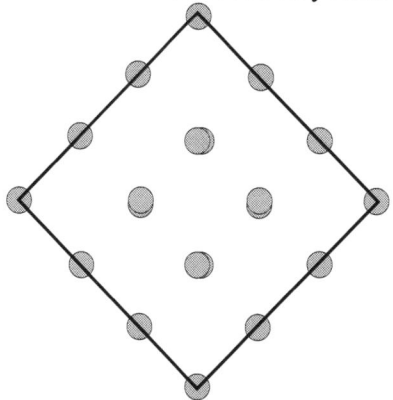

Octavia smiled but said nothing. Anyway, I was ready to impress her with my knowledge of the place.

"I would like to meet some of your clients and maybe visit the council," I said.

"My clients are here. Go ahead and mingle if you like."

I strained my sight in every direction. I could see the entire floor through the mist but did not see anybody. There were no strong guys making structures by climbing on each other shoulders. Upsetting this may have been, but I did not want to be picky with Octavia. Not only that, I knew being picky was not going to be helpful or useful. Think, I said to myself. Better yet, observe.

I could hear Inertia's music but it sounded like a buzz and I faintly saw the mist making some kind of resonating patterns in the room. Patterns were not flat but were in space, all over, and they were all symmetrical from the center of the room. There was then a loud bang, a kind of a slap or a ding of a gong, I could not tell. The entire room nudged to one side and the air cleared up as the mist suddenly disappeared. The first thing that came to my mind was that the strong guys had something to do with the shudder because the room actually moved. I walked over to the gap and the gap was still there, as wide as before. Perhaps we were hit by something but the pizza was still there in the same spot. G had something to do with this, I though, as I recalled Isotopia's ready-or-not comment.

"Your clients are at the council," I said as I watched Octavia smile. Then it dawned on me that I was not at the council and this was something that did not work.

"You are not at the council, Octavia?" I asked, keeping my chin up by trying to put her down. Octavia did not answer and I guessed I was the one to answer my question. It was back to the drawing board for me and I could feel the pain. But Octavia started misting the room and it felt good. Inertia's music was also to be heard. I took a bite of the pizza and was ready to go again. To my surprise, Octavia started to speak.

"It is not easy to mingle with the strong guys and it comes with a commitment as well. You may want to visit the council galleries first. I'll find the gal who knows how to work the galleries."

I liked this witch. I did not have anything to give to her yet she was happy with my progress.

Galleries

"Bleachers are the easiest," said Twiggy. "Let me see how well you can stretch. I could put you a bit closer."

So I made a shape of a doughnut with my hands and stretched it over my head, palms out.

"I'll poke one of the skinny guys out to the bleachers and you can take his place for awhile." Twiggy said. "Follow me closely and wave your hands when going over the gap."

Twiggy had no problem going over the gap. I looked silly flapping my hands as a bird but it worked. Twiggy got between one of the skinny guys and the council seats and this guy just flew way out to the bleachers. Twiggy disappeared and I figured her show was over.

So here I was in the galleries. Nice seats. I say seats because the seat I was in had a view of the council from anywhere in the galleries. Then I thought of my hands being stretched as a doughnut and I noticed my whole body was stretched as a doughnut all around the council and I could see the empty council seats from anywhere and from any part of my body. This was no circus but it was fun just the same.

Then the bang like that of the gavel happened again and the whole council went through the motion again. Council guys were suddenly in their seats, watching each other. I noticed there were the orange and blue guys there. The blue guys were from Isotopia because they were spinning. They were just as strong as Octavia's orange guys and they kept G's center platform straight and level. The orange guys made a great pattern with their bodies that were over and above and below the platform as well. Just as everybody saw that everybody was in their place and that the council pattern was holding strong, the orange guys bolted, flying away from the center. Great jet streams washed over the blue guys and they too, took off. As soon as the council guys started to move they disappeared. The only hint left for me to appreciate was the mist. The council seats were empty once again and I knew I saw as much as I needed to see. Twiggy slung me out to the bleachers and from there I moved easily to the square house.

"Twiggy," I said, "I am glad you took me to the gallery. But I did not see any triangles there."

"Square house people do not understands the triangle house very well – nor do the triangle house people understand the square house. The gap is there to always keep the square and the triangle separated. When the guys fly over here they still think they never left the triangle. I am the only one who can go easily across the gap and even go to any floor but even at that I do not understand everything."

As confusing as it was, Twiggy was actually quite revealing. The patterns, I whispered to myself as I was on my way to Octavia. The triangle house had only the third floor and it was real all right but the triangle was logical.

Into the Mist

I made a special pizza for Octavia with twenty slices of pepperoni arranged just the way she likes it – and I made it in the extra deep square dish. She loved it. She passed over the pepperoni as being too spicy, but that was something for me to work on another time.

I stepped into the mist and let the din come to me. After a while I was in the middle of a party. Everybody was there but it did not seem crowded. It took me a while to find Inertia, for indeed she was everywhere and everybody wanted to talk to her. She said she talks in music because it is the richest way of making conversation. I said I liked her patter and she laughed.

I knew I was in the flame and I knew how to come out or be at any place I ever wanted. I drew the model of the house. As I was finishing it I had no clear idea who would be happy to receive it. Everybody, I thought, who likes to build – and the few who would like to travel.

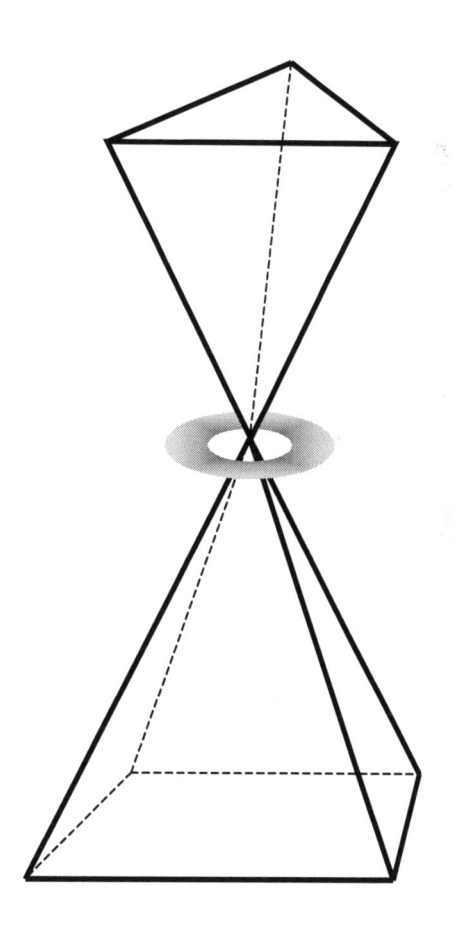

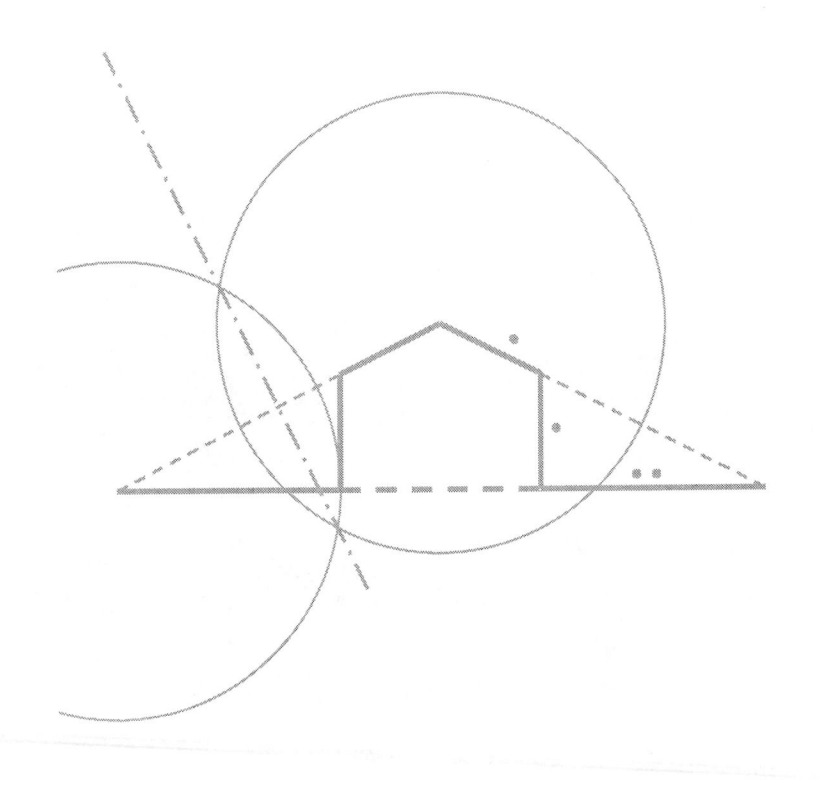

THE TOOLSET	286
THIRD BODY	287
HYDROGEN RECYCLING	288
TIME	290
REAL METHODS	294
VIRTUAL METHODS	297
WORKING THE REAL AND THE VIRTUAL	300
CONSCIOUSNESS AND RECONCILIATION	303
THE OBJECTIVE AND THE TWO POINTS OF BALANCE	309

• • • • Move The Galaxy

> *"Give me a point and I will move the earth!"*
> Cry of Archimedes
>
> *"If you cannot move it, organize it!"*
> Bath revelation, unnumbered

The real and the virtual domains each have their indigenous methods. The interoperability then deals with the separation and joining of the two domains as a way toward increasing organization. The real and the virtual domains are the actual and the existing duality of all there is.

Gauss answered Kepler's musings about the construction of geometric stars in a couple of hundred years. While the Cry of Archimedes remains unanswered for 2300 years, the answers are attached every which way in response to your questions. Indeed, there is a point that moves the earth and the ship. It is not a physical fixed point but it is a point nonetheless, a fulcrum, about which force arises as a result of the addition of infinite virtual components. It is certain that the force relates to the Golden Proportion. The actual mechanism of the infinite and tractable addition is not hidden to those who take the queen that is geometry strolling through the pyramid.

The Toolset

    Reduction is a term with many meanings; one can start with cream sauce reducing over low heat or a reduction in product defects or reduction in fever, inflation or extra harmonics, or anything that at the end of the day ends up with less of something. In this pursuit we are dealing not only with the reduction in chaos, say, but in creations that are concurrent with reduction and where the reduction and creation are inseparable. The early days of alchemy created certain paths and methods that had more to do with transformations than with putting things together or taking things apart. One of such goals is reduction, which seeks new stable states by increasing the quantity and the range of interacting variables. Alchemists always reveled in their ideas and to obtain something new via reduction by putting *more* things into a flask must have been one of their favorite mercurial discussions. Lowering the number of interacting things, the alchemist would say, does not create anything new, and so exclusion and reduction are completely disparate ways of thinking. Adding different things into a pot and thus adding more relationships is how reduction can begin to happen. To increase the range of relationships between things, just add heat.

    Two groups of methods or processes will be separately applicable in the real and the virtual domains. Additional tools draw on mathematical group theory and on transformations that by definition transform some variables – while other variables remain invariant. Time has no role to play. Yet, and just like the alchemist, we will add time to our discussion in a later chapter to see why it does not fit.

    Reduction in chaos has certain tradeoffs. Everybody with a claim to decrease chaos could be right and just about anyone could construct a non-chaotic system. But what also needs to happen is that such stable system is also required to grow because your job as the creator is never done. Application of some of the tools in the toolset will increase the degree of organization and, because the growth in organization is unbounded, the tools will be unbounded as well.

Third Body

Having a general solution means that there are mathematical solutions within broad and practical contexts. Practical solutions include tractable solutions, for mathematical descriptions of reality also happen in real-time. Two-body gravitational interactions are always tractable but the introduction of a third body's gravitational influence breaks up tractability.

In the absence of a formula the path of three bodies can still be simulated on a computer but the simulation always trades off accuracy for tractability. The inability to obtain a general mathematical solution of three or more bodies takes an interesting turn because three bodies can have more than one computable solution – that is, three bodies can be had in more than one hyperstate. It is apparent that if three or more bodies were to become tractable, they need to have a topology that fits a one-body or two-body solution. Taking Sun, Earth and our Moon in the solar context, Earth and Moon become a single body and this one body forms a two-body system with Sun. In another context the Sun, Mercury, and Venus form two two-body systems in a compound fashion. Yet another tractable way to arrange three bodies is that of the Sun, Venus and Earth where the Venus-Earth orbitals interlock thus merging Venus and Earth into one body. Three bodies can also be computable when separated by large distances such that each body is computable on its own because the gravitational forces become nil. A one-body solution calls for merging of three bodies that results in a single cumulative body. Mass size, distance, and momentum all have a role to play in any one particular solution. Chaotic movement of multiple bodies, then, represents a superposition of all possible solutions. The general solution does not exist. This is not because a supreme being or a nasty demon taunts humans with chaos, but because a palette of solutions is available. If the size, distances, and the dynamics between bodies have certain proportions a specific solution can be had. The knowledge of such proportions, then, becomes the most potent ingredient in reaching a solution. Reduction in chaos or a reduction in potential chaos then also converges toward one particular computable solution or a hyperstate out of several available hyperstates. To effect a particular solution, the ability to manage force, mass, and distance becomes the necessary talent in creating computable systems.

287

●●●●●○ Move The Galaxy

Tools and mechanisms for creating computable real systems include the precipitation or realization of matter at the granularity of the hydrogen atom, an enhanced definition of mass inertia, conservation of real and virtual energy, and nonlocal creation of linear and angular momentum in the framework of momentum conservation. You may recall the first principle where no physical or logical variable has a preordained attribute of independence. The mechanism for a sun going nova limits the size of a one-body configuration and it is also a part of the toolset. All real atomic matter remains matter only when it is tractable.

Hydrogen Recycling

The primary mechanism of matter accumulation as a single body is that of matter realization – that is, atoms of hydrogen precipitate inside the body or in space. Free electrons spread rapidly and transit readily into the virtual domain. Spread electrons form ether in the absence of energy, make formations in response to vivibrations, move around as virtual mass, or form hydrogen with protons. Protons also issue from the eye of the galaxy, also in the virtual form. In the eye, atomic compositions that are larger than a proton also transition to virtual and are disassembled into individual virtual protons and electrons through some mechanism. The galactic eye at the center of the galaxy frees matter from gravitational forces there. The absence of gravitation causes matter to become virtual matter, but atomic cores still hold together logically. The dissolution of matter into basic components is then an additional mechanism, consuming very large amounts of energy. Reversing the process later on, nuclear fusion releases the energy as photons when cores become more complex in the solar furnace. It can be said that the collective knowledge of the galaxy takes complex cores apart in the eye by putting in virtual energy. A failed solar system that had undergone supernova may descend into the galactic core at a more rapid rate if it lost much of the angular momentum in the explosion. A sun supplies but a small fraction of the overall angular momentum in a solar system and it is more likely a failed solar system just limps on, surrounded by the dust that used to be the sun.

Quantum Pythagoreans

Hydrogen precipitates in a computable one-body context inside the sun or in free space where distance dominates in all three degrees and provides a computable environment of another hyperstate. The overall mechanism is that matter continuously recycles on the galactic level; moving towards the center of the galaxy as real matter of organized solar systems and moving out as virtual matter composed of the smallest possible building components of electrons and protons that precipitate in its simplest hydrogen form on its way out in the first computable context it finds. Recycling of matter in this way allows for repair and growth of the galaxy. At the edges of the galaxy matter can and does precipitate as individual atoms of hydrogen because such atoms were not needed for repair, and hydrogen materialization then accounts for growth.

Interstellar space contains several atoms of hydrogen in each cubic foot. It appears that their atomic density is such that free atoms do not mutually interact. A sun is a hyperstate of its own and is capable of drawing large amounts of hydrogen because accumulating matter is computable below some density threshold. Few realize that a sun's radiation output would decrease dramatically over a short span of a decade if it were to have but a fixed amount of fusion-able material of hydrogen. A passive system's heat radiates out at the most extraordinary rate proportional to the fourth power of its absolute temperature. Without continuous hydrogen stoking, the sun's output would diminish rapidly. It is then apparent that astrophysicists are reluctant to make the calculations of a rapid decline in sun's radiation using but a fixed-fuel presumption because they have no ready-made answers why there is no fuel shortage. Certain suns are in an oscillatory state where their brightness fluctuates over several degrees of magnitude inside a decade.

Whether hydrogen precipitates inside other massive bodies such as a planet is a topic for discussion. Rather than a new mechanism that continuously creates crude oil and other "organic" compounds, hydrogen precipitation depends fundamentally on the computational context. It is likely Saturn was at one time much larger and was possibly our second sun of a dual sun system. If there ever was a need to decrease Oss angular momentum then this also would have decreased the need for Saturn's mass, and hydrogen burned off, leaving no significant fusion activity on Saturn. Hydrogen may not precipitate inside Saturn at all if the computational result calls for a decrease in mass, but this mechanism is reversible. However, as the number of planets increases it becomes less likely for former second suns such as Saturn to have a significant fusion activity again. Planets hold 98 percent of Oss angular momentum and precipitation of hydrogen in all planetary bodies obviate the need to increase mass in just one body.

Time

Aristotle searched for things causal. The difficulty with causal-only thinking is a continuous buildup of hierarchy because one needs ever-bigger things to prevail over smaller things. Aristotle arrived at a notion of a prime mover that, whether singular or plural, was immovable. It is apparent Aristotle established a model for an organization with centralized hierarchy. Should Aristotle be speaking about time today he could make the best case for time along the following lines:

> "The Prime Mover bidding travelers on their destined journey has the guideposts of time framing the path that never strays. Today with certainty we know what was before and what will follow and time takes its righteous place among the causal elements."

Time, however, is derived from periodicity. Initially derived from lunar and earth's orbits, time is presently referenced through periodicity of atomic decay. Because time is a derivative, time arises always and only from things that are repeatable, lightspeed included. Repeatability establishes the unit of time, be it a year or a million years. Without repetition, time has no units and becomes impossible to define. Time's predictive prowess disappears when a system becomes chaotic because chaotic systems have no repeating or periodic behavior. The inclusion of time, even from another tractable time reference, cannot help in the emulation of chaotic systems. The lack of emulation then precludes the ability to make corrections to chaotic systems using time alone.

Time can be visualized as a shadow being cast – that is, derived – by an object. A shadow will move to reflect the movement of the object but it is not possible to move the shadow and by such action move the object.

A shadow always depends on the position of the object and the shadow is the derivative of the object's position. Time is always a derived, and therefore dependent, parameter and time disappears altogether during physical collisions and at quantum mechanical reduction. Disappearance of time should be expected because instant action does not require time. Three-body chaotic interactions do not have a general mathematical solution and the absence of a solution also hints at the fact that time, or waiting, in and of itself cannot reduce chaos. Time, therefore, cannot in and of itself elicit a solution and time cannot be considered to have causal characteristics.

During collisions between physical bodies momentum is exchanged in an instant and time becomes irrelevant as time becomes zero. Time does not dictate or prevail upon any other variables during physical collisions because nothing waits for time and time is not needed. While it is true, for instance, that one need only wait a while before the baseball season starts again, this takes place in a system that is already organized where a derivative such as time can be relied on. Quantum mechanical creation of momentum that manifests as gravitation deals with synchronization because linear and angular momentum is imparted instantly on two bodies across some distance. Synchronization does not deal with time and in fact the ability to make the event instantaneous discards time altogether. Synchronization is a natural and inevitable outcome issuing from the even symmetry of any and all forces, which act across distance in equal measure. When a photon enters the beam-branching mirror, for example, one can describe photon branching as happening "at the same time." However, a photon parts and branches instantly into two branches *whenever* light arrives at the beam-branching mirror. The evenness of all force wavefunctions guarantees that when force manifests it always happens in the 'equal and opposite' framework that is independent of time.

Gravitational universal synchronization, on the other hand, results from a superposition of near infinite quantity of all possible orbital periods. Gravitational timing that permeates the universe can be visualized as the convergent limit of all possible orbits, which, in turn, are created from all repeating frequencies carried by all rational numbers. Gravitational universal synchronization can then be likened to a time period that finds an

291

agreement, a common denominator, or a unitary multiple with all orbital frequencies. Gravitational universal synchronization is likely the universal time Newton postulated as existing even though this particular absolute time is constructed through the creation of all real entities relying on orbits for their existence. Orbits are central to the macro or the cosmic universe while electron orbitals are central to the atomic universe. Electron orbitals are not associated with the gravitational universal synchronization. The atomic core appears to mediate the two manifestations.

Another argument can be had that one only needs to wait for the gravitational momentum to appear. This is indeed the case. Time is the enabling or gating or synchronizing agent in the creation of the gravitational momentum. The magnitude of the gravitational timing period, however, depends on other variables. Gravitational period or frequency is yet to be worked and substantiated mathematically but in principle it rests on combining all of the harmonics carried by all rational sub-unity fractions that are the frequencies associated with repeating rational numbers. Time is inherent in the constancy of lightspeed but it is the constant lightspeed that creates time and not the other way around. Gravitational wavefunction reduces periodically but this very period is derived from all orbits, which, in turn, is computationally derived from integers and their ratios. The question needing resolution is whether the derivation of the timing period applies to all integers or only to ratios of particular integers.

The upside is that vibrations can be applied in a way that will affect gravitation. Instead of presuming that the gravitational periodic synchronizations are fixed, unchangeable, and "supreme," gravitational timing can be worked by focusing on variables from which such timing is derived. Gravitational timing, then, can be suspended – at least on the local level – and the gravitation neutralized.

There are situations where some event is "just a matter of time." This context happens in a closed system where entropy is increasing and organization is decreasing. Barriers are put in place in such cases as laying a siege to a castle, yet time is but a passive – that is derived – component because the desired event depends on the system becoming and remaining a closed system. The desired outcome for either side depends on things

Quantum Pythagoreans

other than time – and a siege is not successful once the system's closure fails to hold. Time by itself cannot dictate or cause things to happen.

Time is derived from something that has periodic or constant behavior. Absolute time derives directly from the absolute speed of light. Even though time is a derivative and is always a dependent variable, absolute time can be constructed if time is derived from some entity's absolute parameters. Absolute spatial distance and absolute time must be, and indeed were, constructed for the real system to hold and operate as the formal system. If a planet started to deviate from its orbit, one would look for many possible reasons but time cannot be the cause of the deviation. Time cannot be an independent variable because no variable subordinates itself to time.

Should one influence the computational structure of the atom such that the periodicity of the atomic fission changes somewhat – and this can be done with excursions in pressure or temperature – then the time "standard" derived from such atom will change as well. A slow running watch cannot increase our lifespan because we expend a certain amount of energy doing our work and our metabolic processes follow that, regardless of the position of the hands on the watch in our pocket. If every atom's periodicity in a human body changed, the metabolic energy output would still be consistent with the work the body performs and the quantity of time tics would simply become different while the work the body performs remained the same.

An apple falling toward earth is accelerating as it moves faster and faster toward earth. A system comprised of both the earth and the apple is not accelerating, however, because momentum is created only in the framework of momentum conservation and the momentum the apple is acquiring is always the same and always in the opposite direction to that of the momentum earth is acquiring. In a system where total momentum is zero there is no net movement and such a system needs no time to describe it. Fundamentally, time arises as a parameter only if a body or a system under observation is moving. Time becomes a useful – that is predictive – parameter only if a body is in hyperstate. While all hyperstates are by definition tractable, some solutions have periodicity as an additional property. A parabolic path is not periodic while an elliptical path is.

Periodicity creates a time reference, so the first atom created also created time. Time reference is about periodicity and periodicity is contained within a unique subset of hyperstates. When a force claims one degree of independence the planetary orbits are or become interlocked, which is a source of time reference. When force dominance approaches three degrees of independence the orbits of bodies are no longer in one plane and orbits become, by definition, three-dimensional. A spherical galaxy has but one time reference among all solar systems because all solar systems move in synchrony even though the synchronous movement is predominantly radial rather than orbital. Force is the dominant parameter facilitating the creation of periodicity and all hyperstates surrounding hyperstate 3,0,0 have periodicity. Hyperstate 1,2,0 also has periodicity when two bodies orbit each other. With instruments, time can be derived from all hyperstates where force claims one or more degrees of independence. This should be expected if time is identified with motion, for it is force that puts things in motion.

Time can be found in various relationships but in each and every case time is unique to the mechanism that produces it. At times, time is visualized as an arrow but, depending on the context, such an arrow can be made to point forward, backward, be discontinuous, non-linear, become zero, or be a combination of these. Time cannot be removed or disconnected from the phenomenon that produces it and time, in and by itself, cannot be generalized.

Real Methods

Real methods are applicable in the real domain where the role of context is fixed. The primary method is inference in the form of the *if-then* operation because in the real domain this method yields correct results without regard to additional context. The conditional *if-then* operation is the central property of the present day computer. Should a particular *if-then* operation hold, then causality arises because one can rely on this logical sequence to hold in the future.

Quantum Pythagoreans

There is just one result allowed during the *if-then* process. The answer to the conditional *if* is either yes or no and then there is but one action or one sequence of unique actions associated with each the yes or the no. The *if-then* procedure is always about action because 'maybe' is not allowed. Fundamentally, the entire *if-then* process hinges on prior knowledge. Conditional *if* requires the answer to the supposition to exist and the inference *then* also requires existing knowledge. On the present day computer, fetching the answer to the supposition *if* relies on knowing the location where the information is stored and the location is supplied either directly or by searching. The search, however, can be tractable or intractable. In either case, the reliance on preexisting knowledge does not avail the *if-then* process to be self-correcting, self-improving or self-organizing. The *if-then* process always relies on prior knowledge. This knowledge must also be readily available because an intractable search for the answer would not be useful; it would put the answer at some unknown or impractical point in the future.

All real methods treat the procedure as primary while the information, or data, is subordinated or secondary. Methodologies are, therefore, in charge and data follows, for procedures are unchanging while the data is the entity that is being manipulated. Overall, the *if-then* process can also be described as "working the plan," for the procedure is in place and the executing entity always knows what to do with freely varying data.

Real methods yield useful results because real things have finite and repeatable properties. In addition to computability, repeatability includes both the constancy and periodicity. Constancy and periodicity are the two major attributes of the real domain. Real objects cannot have an infinite number of properties because the *if-then* method could not describe such objects tractably. Similarly, the real domain cannot have an infinite number of objects. The best the real domain can claim is the ability to be unbounded or unlimited, but the real domain cannot become infinite. The real percept, then, searches for determinism and repeatability. Determinism is another word for tractability and this attribute includes measurement as one of the 'ways and means' because the act of measurement results in one answer that dovetails nicely with the *if-then* method. Real methods tend toward measurement because real variables provide one answer that is the magnitude and magnitude yields a sufficient and absolute measurement in

295

the real domain. Also, the discoveries of equations that determine future states of real objects reduce the need for continuous measurement.

If object partitioning or object cutting does not change the context then such operations are amenable to real methods. An apple cut in half is just two halves of an apple. Context does not change when linkages between things are few in number so linkages can be ignored. A thousand-piece car engine can be taken apart and then repeatably put together and made good again. A broken off handle of a cup can be made good again provided the cup has but a utilitarian purpose – that is, the cup's link to its collectible value is low and can be ignored. Partitioning things into components or building blocks can be unambiguously and tractably described with *if-then* methods. Similarly, assembly of bigger things from building blocks can also be documented with *if-then* methods.

When the perception of reality dominates, real methods are called upon to explain the workings of the universe. If there were such a thing as the real-only person, this person would likely believe in an inexhaustible supply of real energy because this construct allows the creation of a repeatable and constant environment. A closed system construct also enhances repeatability because a closed system enables the application of stochastic or averaging or statistical methods to establish rules and laws, and rules and laws are none other than *if-then* methods. Strict hierarchy would also make the list of the real-only person because unchanging hierarchy yields repeatability. Because the *if-then* method is independent of context it can then be valid regardless of context: The real-only person classifies relationships as being valid for all – a behavior that is wrong is always wrong and there are no exceptions, for exceptions could bring about a change in context.

Nonpolynomial properties arise when the increase in a system's membership causes the new system's description to require more operations than the number of members raised to a power of some number. In such a situation the computer simply cannot keep up with the increase in the system's complexity – even if many machines are working in parallel – because the computing load increases faster than computing resources. Code breaking is an example of a nonpolynomial problem where the goal is to substitute every coded letter with every available letter until the group of letters begins to make sense. Every time a new letter is added to the analysis a complete re-computing of all possible letters is most likely required. Computational need rises so rapidly that if you desire a result in a practical timeframe a million or billion or whatever number of computers working in parallel does not help much. The apparent conclusion is that code breaking or the organization of a billion-body galaxy calls for something qualitatively different.

The existence of a mathematical solution is the same thing as saying that the solved problem is polynomial and therefore tractable. This holds for knowing the decryption code or knowing the mathematical function that describes the movement of a physical body.

While there are ten major and realizable hyperstates, the hyperstates plane can have a very large number of solutions since each point on the hyperstates plane is a potential computable solution.

Virtual Methods

The Pure State of Maybe

On the periphery of your eye the fairies may be
In the pure state of maybe nobody can catch them
      but you can dance them, altogether swirl all of them, sweet as honey be

At the top of the stairs you will know who si who and what is hwat
Magician weaves the fabric of the craft
He will not create your desire
      You are the keeper of your fire
You cannot be an object of His ire

He will, for sure, answer questions you aspire

Virtual methods are about "planning the work." Because the real method "works the plan," there is a hyperbolic reversal here and the expectation is that the procedure and data will reverse their independent and dependent quality. Indeed, data become independent and thus have superior or leading positions. Procedures and processes now take a subordinated role. The virtual method is about relevancy and can be described as the *what else* method.

What else method is an associative search coupled with relevancy. Instead of sequentially searching the database for data that meets certain criteria, an associative search gives data itself the leading role. If the entity is 'submarine,' all associated data deals not only with things that live or work under water, but includes such finds as submarine sandwich, floods, ocean currents, bubbles, even corrosion. The associative search compiles entities by asking, "*What else* is there that may be relevant to submarine?" While the search itself has no restrictions, the ability to measure relevancy becomes the principal activity after the collection. There are, then, two phases. The first collects all entities associated with locus 'submarine' – wherever such entities may be. The second phase measures the association relevancy to allow prioritizing of all possible answers. The good news is that the speed of data collection can be concurrent if all of entities' data exist in superposition. Looking back at the real methods, when data is subordinated to procedures then each piece of data has its own location with a unique address. Virtual methodology, however, requires all data to be superposed – that is, all variables exist together alongside and on top of each other and there is no such thing as data location. There is no limit to data superposition and the degree of search concurrency is therefore unbounded. Technically, the virtual search method compares something to everything in one operation. All relevant entities then collect around the locus entity in a way that is oftentimes visualized as a vortex.

Each entity is represented by attributes. When an entity's attributes change as a result of influences they are no longer constant and become variables. Variables, in turn, are described with their data values. Virtual methods deal with variables because virtual methods determine the degree of influence among variables. Variables consist of data values that are changing as a result of other variables' changes and in general the virtual methods are about changes. It is most convenient to record all of variables' values as time goes on because this allows a record of all variables to exist as time goes on and, in reverse, a time record allows all of variables' values to be reconstructed. Yet there is more to it than just convenience. When determining the degree of influence variables have on each other, it is not possible to do so with only the values variables have at the present time. The influence and consequent movement variables have can be quantified only after obtaining many mutual readings and this requires storage of data. How accurate a clock is being used for sampling makes but a subjective difference. As long as the time reference is the same for all measurements, an exact reconstruction of the past values of all variables can be made. However, if the absolute clock is used for the recording, then any and all observers will also be able to reconstruct the past values to everybody's mutual agreement. An absolute clock with periodic sampling, then, allows the synchronous storage of variables' unambiguous values. Time periods must be used, for if time were continuous then data storage would be infinite for any particular length of time.

The relevancy determination portion of the virtual method is a separate step from the mechanism of the associative collection. The relevance measurement is between variables themselves because relevance is about leadership measurement and, therefore, variable's leadership can be determined by comparing it with all other collected variables. If the objective is to find variables wielding the largest influence such as when looking for top ten influences, each and every variable needs to be processed with every other variable. Because the leadership quality depends on context, any change in context will reflect in measured leadership values. Different contexts are represented and processed within particular time periods because identical – that is synchronous – sampling allows synchronization and bounding of all variables to an identical context. Data having priority over procedures also means that it is the change in data values that is of primary interest. Variables that do not change are by definition constants and do not enter into the measurement of influences. If some attributes continue to have the same or similar data value in spite of the change in context then they probably belong to a real thing and may be excluded from virtual methods.

Placing another entity in the locus may increase or decrease the relevant collected subset. The associated data may increase if new associations are uncovered. Switching the locus from 'submarine' to 'sea horse' may appear to shrink all submarine things to a family of certain sea creatures, but the associations with horse brings in a new and large amount of associations that will also need to be processed for relevancy.

Leadership is relevancy quantified. To quantify a variable's leading or following qualities, a measure of randomness is useful. Variables that behave randomly can be excluded because they cannot have leading or following qualities with respect to the variable in locus. Large numbers of variables can be discarded after collection if they behave randomly. If the variable is not random, then it can be found to move in phase or out of phase with variables that are relevant to the locus entity and, subsequently, the supporting or detracting relationship can be quantified. Leading and following quantification then also results in the direction of influence or in the determination of a priority. Consequently, at the end of the analysis just described, the determination can be made that with the increase in outdoor temperature the consumption of ice cream goes up but not the other way around. Leading and following variables can be also called dominant and recessive variables. In a similar way, economists talk about 'leading' variables while biologists refer to 'dominant' variables.

Action is altogether absent from the *what else* method.

The virtual method results in a list of all relevant entities prioritized according to the degree of their relevancy where such relevancy changes with context. Context also includes the period under consideration, which is also the number of samples of each variable.

The *what else* method is about joining because there is a continuous buildup of entities that have some relevancy to the locus. Cutting or partitioning works against the *what else* method because cutting severs linkages and relationships. It may be easy and agreeable to cut an apple but a parting of ways after a break up of human relationship is another thing again. Joining is a result of virtual methods while the assembly or disassembly results from real methods. The measure of relevance of linkages is the differentiator between the two. The alchemist would say that wood furniture is joined rather than assembled.

Working the Real and the Virtual

Both the real and virtual domains deal with disparate properties of the universe and have methods that are most advantageous and therefore indigenous to only one domain. Real methods yield predictability by discovering constancy and periodicity while virtual methods process changes that hone in on the most relevant context created by the most influential variables. Real methods by themselves cannot increase the system's organization but they do leverage existing knowledge. Virtual methods by themselves do not perform action but they can reveal the most promising opportunities coalescing from among unbounded relationships.

Real processes *apply* leading variables where each leading variable has a plurality of variables that follow. Real methods, then, are 'one-to-many' processes. On the other hand, the virtual methods identify relevant variables, both leading and following, from among the infinite number of variables that work together toward some outcome.

Another connection to alchemy is in that the King wields the 'one-to-many' processes while the Queen is fond of the 'many-to-one' process. You may come across an alchemical text with the king and queen and a short citation:

> *King makes a proclamation*
> *while*
> *Queen hears the echo of the nation*

You just will be able to enjoy the way alchemy works.

The real and virtual domains are exclusive to each other and collectively comprise the fundamental duality of nature. There can be no possible melding of the two since action and non-action cannot be merged, just as fixed and changing cannot become a simultaneous objective, and because cutting and joining are at cross-purposes. Transition between the real and virtual is through the hyperbolic reversal. Even though the separation is vital to organization and self-organization, one can speak of the real and virtual as being separated by a narrow gap because, while a gap separates, a narrow gap can also be bridged.

Separation of the real and virtual may yield two disparate outcomes. The reconciliation of the two outcomes, however, leads to an increase in organization. Real domain's outcome leverages existing knowledge but if such knowledge is flawed then the leverage will operate in a detrimental direction. The outcome of virtual methods enables a new set of real processes to realize action in a new and likely positive direction. Change for the sake of change will not yield improvement if you do not apply the virtual methods first. Although there is no inherent fixed hierarchy within the virtual method, a new organization chart or a new process flow can be constructed in agreement with the prioritized list obtained by virtual methods. Overall, the improvement in organization results from an ongoing comparison of results obtained separately by the real and virtual perspectives. Because the comparison may at times be synergistic and at other times conflicting, one can speak of the engagement of the real and virtual methods or the engagement of knowledge from each domain. Since the engagement context is unbounded, organization or self-organization is unbounded as well.

The present day computer rapidly executes real *if-then* methods. *What else* methods require all data to be superimposed on top of each other. Data lacks specific location not because it is forbidden but because the access to data is associative – that is, data is accessed *by content* and not by location. Light and a virtual electron have superposition properties that are unbounded and both light and virtual electrons are technically suitable to deal with the unbounded concurrency required for virtual methods.

Mathematical group theory allows the manipulation of particular groupings of parameters associated with some entity. Group theory establishes operations that transform one group of parameters into another while upholding the integrity of the entity. A soccer team is an entity that has defensive and offensive groups. Players can belong to either group and the operation that transforms some players from defense to offense changes the behavior of the group(s) but nevertheless the operation upholds the integrity of the team as a whole. The benefit of group operations is that some group parameters may be transformed and other parameters preserved while the integrity of the entity holds. In the case of the atomic particle, a group parameter called charge has the state of 'positive' or 'negative' or 'neutral' and under a particular operation the charge may be preserved while another group parameter called spin can be transformed from 'up' to 'down' under the same operation. Oftentimes a parameter that takes on only two states is called parity. An electron's ability to become a virtual electron adds a new 'real' or 'virtual' parity to the electron's parameter list that may be preserved in some operations and transformed in other operations while the electron, as an entity, does not disappear. Pythagoreans expended considerable effort qualifying the properties of numbers and were first to establish the first two groups – those of even and odd numbers. Pythagoreans thought of even numbers as feminine and odd numbers as masculine, and also developed even-odd transformation rules for addition, multiplication and subtraction. Apparently, Pythagoreans felt that even and odd parity of numbers carries much significance. Today's computer memories use even and odd parity with 'exclusive or' transformation to detect and correct errors but, more importantly, in the 1840s the even and odd functions were called as such because of their unique properties. Even functions are invariant under vertical symmetry transformation. Vertical symmetry has all even functions appear as reflected in a mirror that touches the vertical axis. Odd functions are invariant for symmetry transformation about the point-of-origin and so the trace of the odd functions passes through the origin. If the odd function's positive values are rotated by 180 degrees about the origin, then the function's negative numbers are created. Atomic particles are represented with even or odd functions. Even functions mathematically describe entities that are nonlocal and inclusive, and such entities can overlap (overlay, superpose) each other. Odd functions, however, are localized and exclusive, and such particles displace each other. There are several summing points here: **(1)** Tools such as the group theory are available to manage additional properties of entities; **(2)** Transformation symmetries where some properties change and some remain the same are important in some way; and **(3)** Twenty six hundred year old math may have plenty of new math in it.

Consciousness and Reconciliation

Human brains and the brains of many other species are noticeably divided into two hemispheres. Because the two halves are also joined at the corpus callosum, the riddle of the separation and joining is answered with the real-virtual duality. The brain emulates the world the way the world works and the "joined separation" of the brain is the manifestation of the interoperability of the real and virtual domains.

Both eyes perceive nearly the same scene. Images from both eyes are delivered to both the left and right hemispheres, and each hemisphere responds to the images differently. The left hemisphere responds to real aspects such as unchanging hierarchy while the right hemisphere deals with relationships such as the leading-following strengths happening in particular contexts. The left hemisphere of the brain (the left brain) is stimulated by images conforming to the *if-then* methodology while the right brain is stimulated by the *what else* methodology. Hemispheres do not exist in a state of conflict during the process of image stimulus extraction.

The corpus callosum, by itself or in connection with other structures, is responsible for inverting all hyperbolic functions. The inversion takes any and all hyperbolic relationships in the right brain and symmetrically reflects them into the left-brain. Inversions or reflections about symmetry can be formalized as transformations. Any and all variables that are independent in the left-brain become dependent variables in the right brain, and vice versa.

Consciousness is about representing all stimuli from the environment in an actionable format that can manifest in the best possible context. When there is a dual arousal – that is, dual response – that corresponds to both the real and virtual aspects then there exists consciousness in one's mind. The separation of the real and virtual domains requires that a dual representation of the stimuli is made first. Representation of the environment in the real context is structured for applications applying the *if-then* methods while the virtual representation is suitable for the *what else* methods. Consciousness enables the actionable response of both the real and the virtual methods. Real methods respond with objective actions such as physical movement or rational thought whereas virtual methods respond with subjective action where the relevant context changes. Reconciliation deals with actions and context changes that may find themselves in conflict. Reconciliation can happen later as well and is accompanied by the integration of all previous stimuli and responses. If the conflicting perspectives from the real-virtual processing are resolved, the outcome is continuous self-improvement and self-organization.

When two objects of unequal weight accelerate toward the ground, we could analyze the event as follows. The heavier object is at some ratio to the lighter object such as three to one. Our left-brain cuts the heavier object into three or so pieces in order to obtain pieces of equal weight. That being done, all objects of equal weight are now subject to the same force and each individual piece is accelerating at the same rate. This result is forwarded to the right brain where virtual methods determine that any two or more pieces joined together will also accelerate at the same rate because no object is falling faster or slower than any other object before or after joining. To generalize further, left-brain processing produces the result where no matter what magnitude the gravitational force may increase to, all pieces will fall at a constant rate as long as they are all of the same mass. This step advances 'gravitational force per unit mass results in constant acceleration' to 'any gravitational force per unit mass results in constant acceleration,' which generalizes the force that acts on the object. Gravitational force may be different on different planets but the gravitational force does not change during the fall. A measurement may be in order to verify that a particular object weight is the same on top and at the foot of the tower at Pisa, for example. Indeed, for all practical purposes the force of gravitation does not change with the height of the tower.

When the left brain is engaged in cutting or dividing something, the right brain should not be cognizant of such action because chopping things up confuses *what else* methods. It is the real perspective's job to create things by partitioning. Returning to our example, all objects in the virtual domain are related among each other and joined into different groups but all groups are found to be equal as far as acceleration is concerned. Virtual methods establish that a particular group and an infinite number of other joined groups are qualitatively and quantitatively identical as far as constant acceleration is concerned. The virtual side's job is to relate and to join and make structures robust by putting things together with relevant linkages taken from the pool of the infinite possibilities. The left brain, then, should not be cognizant of the joining action because it cannot deal with infinities. Overall, there must exist isolation and separation between the real and the virtual domains, else the real and virtual perspectives get hopelessly confused. From each other's perspective, and should one side be cognizant what the other side is doing, they would each see it as working at cross-purposes. If the separation did not exist all processes would be working together, confusing each other, and it would be difficult to advance the problem because the advancement requires separate processing *and* transitions between real and virtual perspectives.

Move The Galaxy

It is likely that the left brain must be trained to present problems to the right brain in a format that is rationed-out. Before the right brain can apply its ability to relate things in a potentially infinite context it must have such objects "cut up" in comparable ratios. The problem is first converted to gallon-per-minute or feet-per-second before the right brain can advance the ratios toward a more general solution. In the unequal-weight falling body example, the force acting on any unit of mass is the same because all objects are cut up in equal pieces of mass, and then all pieces accelerate at the same rate. Putting this as an equation, force per any unit of mass = same acceleration rate, or **F/m = a**. This equation is usually written as **F = m·a** and is the hallmark of classical physics.

The rationing of variables has but one purpose: present the rationed entity in a *timeless* fashion – that is, the force on objects that are sliced into equal weight pieces is time invariant. Ratios such as weight-to-length of a pendulum have utility because they are time invariant and can be readily put in locus by the right brain. *What else* methods deal with changes in locus attributes but the *what else* method cannot deal with changes to the locus itself. Right brain deploys infinite concurrency associated with the locus but if the locus itself changes then its associations, relevancy and joining are more difficult to generalize. Time invariance is not about excluding time. Rather, time invariance confirms that the entity as a whole does not change for the purpose of our needs. One can also say that time invariance of the entity in locus allows virtual methods to advance generalization because the integrity of the entity does not change.

The outcome of the left brain reasoning calls for a rationed entity that is time invariant with respect to the objective or with respect to the problem at hand. The outcome of right brain reasoning, "on the other hand," should be such that the left brain is thereafter capable of making the most general inferences in the *if-then* framework. Right brain has done its job well if it presents the left-brain with the widest possible context of applicability – that is, the supposition *for all* is valid. The widest possible context that is valid *for all* then also means the left brain reasoning need not be concerned with context. The real domain deals with laws and rules and is not burdened by context, and is not well suited for context, for context takes the variables into infinities. Overall, then, the presentation of a problem suitable for right brain processing needs to be cohesive (time invariant) while the presentation of a problem suitable for left brain processing needs to be context invariant.

It is said Aristotle claimed heavier or denser objects fall faster and it is certainly the case that the heavier object has a greater impact if it were to land on your foot. Aristotle's context always includes some medium such as air or water and it is then likely he thought of the rock and the feather as always falling in air, for example. Galileo, however, changed the context by removing the air. It is interesting that Galileo analyzed problems as ratios and it is likely he would always write the equation as $F/m = a$. While the $F = m \cdot a$ equation is also in a mathematically correct form, the rationed version should be more pleasing because it is easier to comprehend. When dealing with gravitation, 'force per unit of mass results in (is) constant acceleration' is in superior format to 'force is (equals) unit of mass multiplied by constant acceleration.' When the statement is made in rationed fashion that is time invariant, the right brain immediately jumps in with many associations and that is the reason for easier memorization of the statement.[84] When the statement is made in the procedural fashion, the right brain cannot advance it but the left brain is ready to apply processes where data is subordinated. The statement, 'force equals unit of mass multiplied by constant acceleration' likely elicits the left brain response: 'I measure acceleration and multiply it by mass to get force – and I am done.' Leveraging is a multiplying process well suited for the left brain because multiplication is a component of a given process that manipulates subordinated data to get the job done. Multiplication and division are processes well suited for left brain processing and such results may be the final result for a given problem. If the result is not a final result, the left brain cannot advance the problem and the problem needs to be forwarded to the right brain. If the result needs to advance along right brain reasoning, however, it must first be processed into an entity that is cohesive and, therefore, time invariant.

---

[84] A case can be made that the absence of ratios and fractions from Mayan cosmology, culture, and science was a significant deficiency.

Although the rationed version is easier to memorize, the timeless aspects of the right brain processing elicits no physical action, for it is the left brain that acts. When the right brain creates all possible associations and relevancies with the entity *fish*, one of the context invariant outcomes is that 'all fish are swimmers.' The left brain now has a field day with 'a minnow is a fish and, therefore, a minnow is a swimmer.'

If the right brain does not process its joining and comparing methods very well, the supposition *for all* may be tenuous or erroneous. Take the following statement: "All people have the same metabolism. Jack is a person and drug X helps Jack with Y problem – therefore, drug X helps any person with Y problem." While there is no error in the inference because the conclusion is formally correct, the inference may not hold if some people have different metabolism or if metabolism varies with context. Some person's metabolic chain segment may not work properly and such person should then be excluded to keep the inference valid. Similarly, if a person metabolizes differently while under stress or under unusual regimen then such contexts should also be excluded. These are exceptions that almost always happen but when they begin to accumulate the reconciliation of the conflicts may become intractable if going through too many exceptions, which, in turn, invalidates the *for all* supposition. It is easy to continue to apply 'all birds fly' supposition, for there are but a couple of exceptions, but generalizing the metabolism of each and every person is not possible.

Of some interest are results that have valid hyperbolic reversal. All fish may be swimmers but the reversal where all swimmers are fish does not hold. Valid hyperbolic reversal is a reflection about the diagonal – that is, 45-degree axis – that can be used directly in the other domain. When light waves were thought to have particle properties, de Broglie formulated a reversal where all particles have wave properties. De Broglie's assertion holds for moving particles but, for better or for worse, not all waves behave as particles since light's momentum remains virtual during reflection – that is, light's momentum parity is invariant during reflection (momentum remains virtual) because light does not exert pressure on a mirror. During absorptive interactions of light with matter, however, a photon of light can eject an electron at some velocity and in this case light does behave as if light were a particle because the photon's energy parity transforms and light's virtual energy is converted to real energy of the moving electron. One half of a photon's energy is imparted onto an electron while the other half is imparted on the core – both in the framework of momentum conservation.

The relationship between distance and gravitational force is reversible and so is the hyperbolic relationship between product price and quantity sold. In general, relationships are not reversible and it is necessary that each hemisphere construct the portion of the hyperbole that is in agreement with the real or virtual processing methods.

Although there is no guarantee that the left brain properly delivers time invariant entities or the right brain delivers completely context invariant entities, both the real and the virtual methods are not intractable. It is easy to see that inference is polynomial once the supposition *for all* is found. The unbounded concurrency that is required for the *what else* method of the associated variable extraction is polynomial if data can be superposed without bound in polynomial time.

The Objective And The Two Points of Balance

It is apparent one should maintain some balance between real and virtual methods because the exclusive use of real methods does not increase self-organization and self-improvement. The exclusive use of virtual methods increases the knowledge of our environment but does not lead to a particular action. The second point of balance is between real and virtual domains. Real and virtual domains do not overlap but are bridged by the process of reversing the independence and dependence of variables. Real and virtual domains can be visualized as touching at one point and it is this point that represents the second point of balance. Our mind resides in, and is represented by, the brain. The physical separation between the real and virtual domains of the mind is at the corpus callosum, which is composed of many thin tubular connections. Every tubule is the zero axis that contains the single point of contact between a particular system's real and virtual representations residing in the left and the right brain.

Real and virtual methods can then be engaged to give rise to an autonomous and self-organizing system. External stimuli can be said to result in reactive or self-sustaining measures but, without an objective, the system so far has only a passive existence. The introduction of the objective gives the system an active or proactive bias by first aligning the system with the desired objective or goal. Once processed with virtual methods, the goal modifies all relationships and weighs them to be consistent with the goal. These modifications are purely subjective and one can use terms such as 'the goal has been seeded' or 'the goal has been assimilated.' The goal resides in the first point of balance, the locus, located at the center of the virtual domain. The goal is fully integrated with all relevant systems when it is in balance about the first point and, by definition, the goal has no conflicting relationships in the subjective sense. The virtual domain is at times referred to as the mandala – four cardinal directions centered in a circle, which facilitate the balancing or centering action with respect to the goal. The first point of balance is at the center and placing an entity there, such as the goal, can be described as putting the entity in locus.

The two points of balance are visualized as shown below.

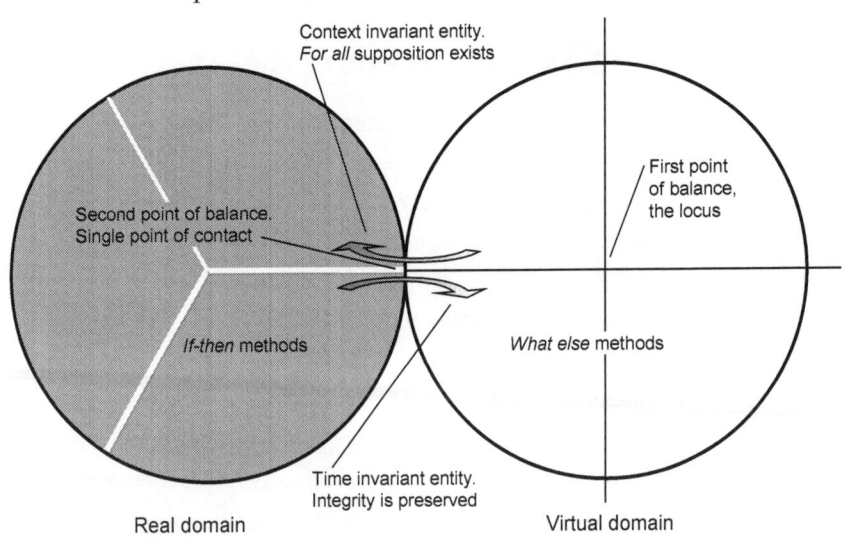

Self-Organizing system

The shape of the human body also conforms to the duality of the real and the virtual. As one may surmise, the balancing actions of the human about the first or the second point will deal with the growth and the actions of the mental and the physical body. If a living body is to intercept

Quantum Pythagoreans

the information available from the real and virtual domains then the human body shape will contain such geometry and will contain the first and the second points of balance as well. The Tao symbol is well suited to represent the second point of balance, for it signifies the interchange dynamics of the differentiated real and virtual domains.

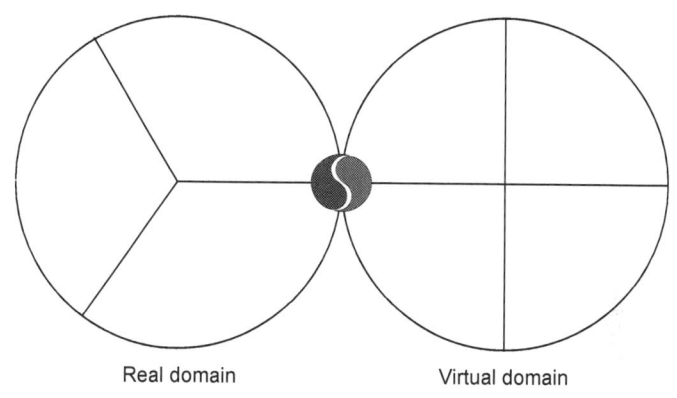

Real domain   Virtual domain

Second point of balance: Dantien or Hara

    In the Eastern tradition this point is well understood and is called dantien in Chinese and hara in Japanese. Dantien consists of two distinct paisley shapes that are closely aligned but do not touch or overlap. The second point of balance has no English word or expression. The word *coupler* would fit nicely as this word describes the engagement of the real and the virtual domain. Technically, the second point of balance is very similar to the optical contact where light's wavelength can couple across a gap. In addition, the mathematical transformation between the real and the virtual domain is accomplished by reflection about the *diagonal*, which facilitates the exchange of the independent and dependent relationships. Moreover, the diagonal row in matrix mathematics has very similar significance for the transformation computations. Putting it all together, the diagonal direction is well endowed in the letter X. The second point of balance can well be called the *couplex*[85] and is shown below.

---

[85] The couplex embodies the transforming and self-organizing function. Consequently, the diagonal direction has a strong symbolic meaning and may well be included in the US flag at some point.

● ● ● ● ● ○ **Move The Galaxy**

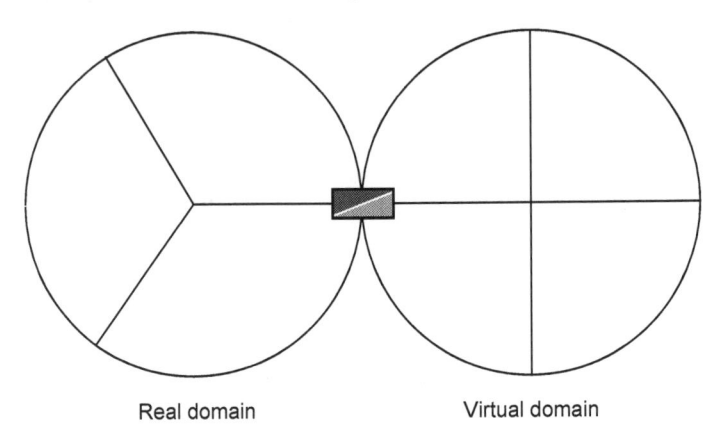

Real domain   Virtual domain

Second point of balance: The Couplex

    In the human body the second point of balance is located at about the height of the navel. Both legs and the portion of the spine up to the couplex have trivet geometry that maps into the real portion of the human body and is closely associated with the body's physical (real) movement. Both arms and the spine above the couplex have quadrature geometry that maps into and symbolizes the virtual portion of the body, which is closely associated with breathing and balance. The outstretched arms and the spine are crossing at the virtual representation of the body and define the first point of balance: the locus or the heart chakra. Movement of the legs is linked to the real movement of the body while breathing is linked to radially symmetrical movement of the body, which is no other than the odd symmetry or symmetry about a point. Balance, on the other hand, is about even symmetry, which is symmetry about the vertical line that is the human spine.

    In the human brain, each thinking system processes information that has its real and virtual representation as well. Thinking systems are touching at one point that is the second point of balance. The corpus callosum is comprised of a plurality of couplex connections that are made between all systems and each system has one interconnect or one point of contact. Each and every system can then improve individually and that is one of the tenets of self-organization. The second point of balance is the point where real and virtual meet, interact, but remain separate. Descartes proposed the pineal gland as being the information interchange center for the entire brain, as he saw a need to arbitrate and reconcile the duality of

Quantum Pythagoreans

the body and the mind. It is likely that the overall first point of balance exists in the pineal gland as an overall superposition of all first points of balance from all right-brain thinking systems. Particularly interesting is Descartes' application of the cross centered system, now called the Cartesian system, that allows relational processing of double-ended variables.

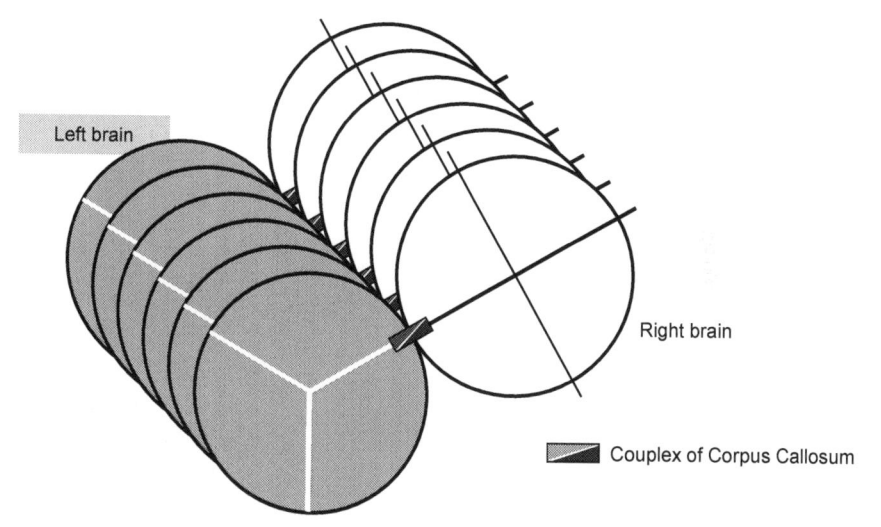

Separation and bridging of self-organizing systems inside the brain

    The left and the right brain emulate the real and the virtual aspect of the universe. Above the "lizard brain" that is driven mostly by lookup tables and interrupts, the cerebellum is split into two halves yet the two halves are joined at the corpus callosum. The brain, then, also contains equivalent representations of the first and the second point of balance.

DAY ONE. KNOWLEDGE AND MOTION	316
DAY TWO. LIGHT	319
DAY THREE. ORGANIZATION	323
DAY FOUR. MEMORY	328
DAY FIVE. TRAVELIN'	333

Genesis

The Magician
Shall not reveal the virtue of the craft
For His thoughts are architect
Look up in His skryes of blue
and see the verse that's born anew
The next the better the last part
if no anger's in your heart

In retrospect it was not difficult to engage the Creator in a conversation. Likely anybody can do it, particularly if you were visiting in his or her seat for a while. All you needed to know that if you challenge the Creator you would be challenged in return and then you had to figure out the challenge.

●●●●●● Genesis

Day One. Knowledge And Motion

We all know the Creator as a being of some stature and statute, and I am glad You came here this week to talk about something You have done for all of us. Creation to me is either complicated or mysterious but in any case what I read about it does not do it for me. Over the years many people claim direct knowledge or privilege of representing You or speaking on Your behalf. Some mysterious entities may even take advantage of our current knowledge and pretend to be You. Today's creation theories are vague or corrupted either by design or by overzealous scribes. Perhaps the creation as we think we know it is but a good story, for it reflects the way we would prefer things to be. But while a story can make you feel good sitting around the fire, it will not get you places. While the fire may keep you warm, the present creation story will keep you Earth-bound. You may have spoken on this topic before but what is needed now is an update, a correction, or maybe a complete discard of the Genesis as we know it. It is likely that some will not see our interview as genuine and so the merits will have to come through the trials, denials, splits, and some old fashioned honest work. For my part, I will touch on as many topics as possible to make the Creation as relevant as I humanly can.

I would like to call you the Magician or perhaps you prefer the Creator, for there are many names people know you by. There is no better place to begin but ... in the beginning...

> "At times people even try to call me a number and – I am listed. I have no strong preference for a name, for one's name is the name of one's deeds.
>
> I could start with space but space has too many meanings and preconceptions. What I really need is spatial distance or just distance. Distance is from here to there because distance is about separation between things. As you can see, I do not even need to create spatial distance because we don't have any planets as yet. Once we have some things, we can use their own size to measure how far apart they are and then the distance will happen by itself. Distance is self-referencing."

Quantum Pythagoreans

Okay. So far you tell me all about things you do not need to create. You are taking it really easy.

> "It is very important to know what I don't have to do. What I said so far is that the universe will be scale-independent – not an easy concept to figure out if you are born into a finished universe. Independence from scale also keeps the universe unbounded because there is always a way to put more distance between things while keeping the same working scales and ratios between things.
>
> What I said is what has already been discovered as commensurable ratios. All real things relate to each other by the measure of other real things. I want you to be alert, but you also want to be in command of the knowledge of your predecessors."

When I see a star, I do not know off-hand if it is a planet, another solar system, or an entire galaxy. You talk about working scales. Are we getting ready to do some work?

> "We are right in the middle of it. I was hoping you could figure out the obvious. If I want to create something or do some work, I will need energy. But there is no energy."

But that's too obvious. You will now create some energy. It's a good start. Do some magic and make some energy out of nothing.

> "No problem, I am also the creator. What are we going to do with this energy? We do not have a ball to get that rolling. We can save it, I guess, in the energy bank until we need it. But since we cannot use it, we do not need to apply it. Better yet, as soon as we will have a need for some real energy, I'll create real energy right away, be assured."

Are you serious?

> "Think."

You did not create energy but you created the potentia for energy

> "Virtual energy, exactly."

To say that we now make up a virtual ball to play a virtual game is not going to fly, you hear?

317

● ● ● ● ● ● Genesis

> "The manifestation. You sure cut to the quick. To manifest something, anything, we will need a real thing. The manifestation will be about movement and to move a real thing we will need real energy. We have virtual energy. Can you convert virtual energy into real energy yourself or do you need magic for that?"

I guess you want me to figure it out. Virtual energy has zero real units. If the real energy equals virtual energy and becomes real, then the real energy must manifest as two things moving in opposite directions because then the total kinetic energy of these two things altogether will be zero.

> "How do you convert real energy back into virtual energy?"

As the two things are moving in the opposite direction, their mutual kinetic energy must decrease ... wait a minute ...

> "Let me make a note on that: 'To conserve real and virtual energy, create gravitation.' Well done. Anything else you want me to create?"

Light!

> "Let's do that tomorrow. Besides, you are not in the dark, really. Meanwhile, think about what needs to happen for real energy to remain real and having no need to be converted back to virtual energy. I know what light is about and you know it takes two to tango."

> **In the beginning all was still and the Magician said onto all: "Move, for motion will manifest your knowledge."**

Day Two. Light

Good morning. Light does not carry real momentum with each photon because light cannot push a mirror. While this is a simple fact, it sets a high challenge to a human intellect.

> "Next to gravitation, light is the most difficult entity to understand. In addition to the separation of the real and the virtual, one also needs to get into the electron. Forces inside the atom are not gravitational but nevertheless there is force there that involves light."

Why complicate things if we don't have to? Why didn't you use gravitational force inside the atom as well? After all, it looks like we are negotiating the exchange of the real and the virtual energy once again?

> "Let me defer that and sum up what happened yesterday. We have the virtual aspect of the universe, which is about knowledge. Then we have the real part, which is about order. Yesterday, the gravitational force worked out nicely as a conduit between the real energy and the virtual energy and now we can treat both the real and the virtual energy as equals, even though each of them has a different purpose."

The knowledge, then, is stored somewhere. You said yesterday we could save virtual energy in the energy bank. I guess it would be a virtual bank?

> "I am glad you are looking forward to this creation and your patience shows that. Yes indeed. I needed storage, lots of storage. Memory on top of memory. All that knowledge has to be here and, also, it needs to be available to everybody and to everything! You can imagine when I said 'Move,' some things did not know how to move and work the energy balance at the same time. The idea is that the universe had to get going and the universe then needs to rework some of its parts while growing as a whole. Well, I knew we all had to start small. Atoms first. I was looking for something that could easily pop in and out of the virtual – something that could be equally at home as a virtual entity or a real particle. I am very proud of the electron now, but at the time the electron was a showstopper. You see, the electron in the virtual domain freely

associates just like any other knowledge and acquires wonderful shapes that reflect its interactions. But the electron also needs to stick around sometimes. The electron has the toughest job because it needs to have inclusive and exclusive properties. The electron needs to relate with memory to access knowledge and also needs to have mass that stands on its own, unchanging. Gravitational force manifests periodically among real things but I needed the electron to be real or virtual as long as necessary. So, when I exempted the electron altogether from the gravitation force, I had to give it the electric charge that stays active at all times and manifests as force at all times.

The electron, in its virtual state, can have a broad range of virtual energy. But making orbitals with electrons cannot have smoothly varying amounts of energy because the orbitals cannot be arbitrarily close to each other."

I know the answer is light, but I do not see why. Electrons do not have the energy conservation issue going from being real to being virtual. What you are also saying is that the orbitals are quantized. So now energy orbitals are spaced out with some minimum difference of energy.

"When an electron transitions from one atomic orbital to another orbital, it does it to perpetuate the atom's exclusive and real existence because electron jumps are in response to external stimuli. An electron's energy level is but a state where the electron is in balance with the inflow and outflow of energy. An electron transitions from real to virtual in its entirety as it spreads around the core and has but the virtual energy of the orbital it is in. Now your answer comes in because light is the excess of the virtual energy an electron has when the electron descends to a lower energy level where such excess energy forms the photon of light. Going the other way, when a photon of light is absorbed between the core and the electron, the electron bounces higher as it spreads and fits another orbital."

You are saying that light is a virtual energy. How come I can see and feel light if light is virtual energy?

"Light is a virtual entity and the fact that light can be superposed without bound is the only clue you need. When light is absorbed it transitions to real energy as vibration that is heat, and that is how we feel it and that is one way light can do real work. If light is not absorbed it propagates as virtual energy and optically relates to matter that may be in its way. In order to actually see light, it must be converted to real energy on the eye's retina – that is, inside matter. Being a virtual entity, light never gets tired going through space. But not even light can be both real and virtual. Each photon of light makes its own transition from virtual to real. What also happens is that the energy is conserved when the photon's virtual energy transforms to real energy.

What is very fundamental here is that no entity can be both real and virtual at the same instance."

Why?

"Real and virtual have different goals, purposes, functions, and parameters. If you combine them or mix them, they will give you mixed up results."

Why?

"Because they interact with the environment in different ways and have different perspectives. They each think differently."

Why?

"Neither the real nor the virtual can figure out all the answers by themselves. But together they can if they are allowed to work on their own. Most of the time, however, real and virtual must be kept apart consciously and actively."

Sounds mysterious. Real and virtual must be separated but they have to be put together to be useful. What puts it all together once the real and the virtual come up with their separate answers?

"For atoms the comparisons of the separate answers must add up to the conservation of energy – both the real and the virtual energy."

And if it does not?

●●●●●●● Genesis

"Matter's real and virtual components separate into matter and antimatter, and we go nova. We then get a lot of photons of light because previously stable – that is computable, real matter converts to energy. The conservation of real and virtual energy prevails even if it means that matter converts to light. The creation of antimatter is a separate topic and it is something I leave to everybody and everything to figure out on their own."

Why is that?

"Antimatter is not something that is easily understood and it deals with reversible and one-way transformations. In many ways, antimatter is unavoidable and at times necessary and at times helpful in a larger context. Matter and antimatter can be healed but in a way it is like a divorce: tough to manage. Having said that, there is no going back once matter is destroyed because both the knowledge and the real thing is gone."

What about tomorrow?

"What about it?"

I thought perhaps we could create planets and orbits and harmonize the whole thing. Personally, I would like to know about superluminal traveling ships — I mean, really, not just imagining

"That may take more than a day. I like the way you put it, though, because first one needs to build things right. Pretensions will not work if you want to do better than imagining."

On the second day the Magician created light and said onto all: "All will see light when matter is created, sustained, or destroyed – and the wise will know the difference."

Day Three. Organization

People try to control their environment and each other through hierarchy or indoctrination. Oftentimes people act in Your name and some do not hesitate to see You not only as the Creator but also as the Commander, Enforcer, Avenger, or Destroyer.

> "It is easy to recognize chaos but it is not easy to diminish it or prevent it. You did not have much difficulty with two bodies because energy can be conserved between the two of them in real and virtual forms and what really helps is that there are definite ways of doing it. With three or more bodies, however, it is not as simple."

Two bodies are easy because two bodies can be interconnected with a single line. Between two bodies we have linear momentum that can be applied with any one degree of freedom and we have angular momentum that can also be applied with one and the same degree of freedom. But the split between linear and angular can happen in an infinity of ways

> "The proportion of the linear and the angular momentum may be unknown right now. We know that each the angular and linear momentum sum up to zero and that is why we need more than one body in a system."

The solution, then, deals with figuring out the proportion. So there must be some context in which the solution is to happen, because linear and angular momentum has two unknowns each and without the other, larger context we have three equations and four unknowns. So you created a third body to give you the additional context.

> "I am losing you. Now, as I recall, I did not need to create the third body. I think it was the other way around. The third body happened when the context was right."

I could have guessed. It happened by itself

> "It did. But you can do better than guess. You can figure it out.
>
> When you worked out the gravitation, you noticed that the slowing down of the two bodies meant that the exchange of the real and the

virtual energy is really energy conservation at work. The energy-conserving mechanism diminishes the real linear momentum that exists between two moving bodies."

You are talking about inertia!

"Am I?"

Could it be that simple?

"What did I say?"

Whoa

"Exciting."

I was trying to figure this out for a while. Why matter offers resistance when one tries to speed it up

"Obvious, isn't it?"

Now it is. If I give matter real kinetic energy, matter can accept it at some rate that is proportional to its own... mass. Matter needs to store as much energy as it is receiving because we know we must conserve it. You see, I know, you know, but matter must also know. Of course, that is why inertia is the same everywhere in the universe. That is also why inertia exists for linear motion and for spinning angular motion

"Well..."

What a tangent. We were talking about the third body and all of a sudden... Okay, what I did was I reversed the scenario. We were talking about the bodies moving away from each other and slowing down because of gravitation. I thought about a body speeding up if I apply force to it.

"Hyperbolic."

Anyway, I can see your point about storage, now. Real energy's value must be stored somewhere and stay there even after the guy who put the energy in is doing something else. Real energy is in kinetic form and it can also increase or decrease subject to computations that transfer the energy between its real and virtual forms because such transfer is in the framework of the conservation of energy.

Quantum Pythagoreans

"So..."

Third body. Yes. With three bodies the momentum can be applied in any two degrees of freedom. Three bodies always comprise a plane. Technically there is no problem in summing their momentum to zero, both linear and angular. The problem, though, is that the number of unknowns goes to six and the number of equations stays at three. So, while momentum is conserved it can also be applied in many different ways.

"Sounds about right. I scratched my head on this for a long time. There are so many different ways of applying momentum among three bodies; bodies can move chaotically while conserving energy. Bodies do not behave in predictable fashion yet momentum and energy are conserved. There is, however, no periodicity and no order. Without order there is no causality or inference or predictability.

Without periodicity the atom cannot sustain itself and so the whole real manifested world can exist only if it is computable."

Are there situations where computability just cannot be had?

"You can always compute. The thing is, can you get at the answer before the answer actually happens? In other words, does a mathematical solution exists that will make the whole embodiment predictable? If bodies disperse and each body conserves energy on its own as it interacts with other bodies, then the situation can become chaotic and no mathematical model is able to describe the interacting behavior of all dispersed bodies in real-time. Bodies become intractable and they are on their own. When the situation is computable then I have an embodiment that is tractable. The challenge, then, is to be able to grow the universe while being tractable at all times – not only somewhere in the universe but also at any arbitrary portion of the universe.

I may be creating things all the time but I also had to point out a bound on degrees of independence. With three degrees of independence the computability – that is tractability – exists but it does not happen automatically and all bodies must be affected to produce a solution. Real entities cannot have more than three

325

> degrees of independence because solutions cannot be had with more than three.
>
> This is a good time to say that you kids have it really good today because there was no time then and to this day there is no way of knowing how long it took to get the third body created just right."

Are we off on another tangent or do we need to create time for this one? Let me say that time will happen by itself because once we get into periodic orbits the time gets nailed right there.

> "I am beginning to enjoy our conversation. You may even figure out that the variables you do not need to create are the dependent variables.
>
> Let's revisit the two bodies with a dash of that inertia of yours. Put them far apart and let them be."

If I create only linear momentum and no angular momentum I have as many equations as unknowns. I take two bodies a large distance away and the bodies speed up toward each other predictably as the linear momentum is realized at both of them. Now, there is no conflict here if we introduce angular momentum but the proportion is still unknown

> "You can suspend it there for later. What is sufficient now is that the introduction of angular momentum comes from a similar atomic mechanism as the linear momentum."

So the angular momentum adds spin to the bodies, and because they rotate in opposite directions the total angular momentum between them adds up to zero and so the creation of spin happens in the framework of momentum conservation. Overall, the created spin slows down the increase in linear speed of the bodies

> "You are just about there. Keep adding spin."

The body splits in two. Nice. Once the body splits while spinning, both pieces continue to orbit each other and their combined angular momentum continues to increase when the bodies have a larger and larger mutual orbit. Each body would also start spinning around its own axis just as the single body was and the bodies would spin and orbit in the same direction. Say, can the two clusters actually slow down rather than just stop increasing their mutual speed?

> "It makes no difference what portion of the total kinetic energy is in linear and in angular form."

Makes sense. We now have three, four, maybe more gravitationally interacting bodies. Because some of them split, we now have them orbiting each other if they are about the same size. If not then the small one orbits the large one. If there are more than two bodies in a cluster, they will orbit in one plane. One result I like is that the proportion of the linear and angular momentum is really not a particular fixed ratio but an interim solution where the ratio can go up or down as necessary.

Now, this works fine if the universe is shrinking. But the universe is expanding. Or is it pulsing? Or are some parts shrinking and some expanding?

> "The universe is organizing. On the second day I got the electron and the atom going along with light. The electron can readily become virtual."

If the electrons and protons move as virtual mass and pop up somewhere at the edges of the universe – well, my, I think I know where we are heading.

> "Not that fast. When the two clusters of bodies are approaching each other, their mutual speed slows down but it is not zero. What I had to do is get the two clusters to begin to orbit each other so they do not crash. Sweet piece of work. You can sleep on it."

On the third day the Magician created a body that moved among and along many other bodies without chaos, and said onto all: "Three real entities compete for three degrees of independence and all by all can find order."

●●●●●● Genesis

Day Four. Memory

We are in our fourth day. What really intrigues me is memory. I was doing some work on electrons and in particular the electron corona that seems to organize in what appear to be one-time learning configurations or geometries. In any case, I agree that any knowledge needs to be stored in memory if we hope for organization to increase.

> "I got the whole thing moving on a very small scale. Atom first. The enforcement of a few simple rules is easy but I could not get atoms to stay stable without much-increased complexity. It took a lot of tuning for atoms to learn how to remain tractable and therefore stable. The solar furnace got too big or hot at times. It was a slow and tedious process to work in all possibilities, the result being that atoms did not grow in size very much but I could replicate the simplest one easily enough. The knowledge we all worked on so hard had to not only stay saved but it also had to move along with the atoms, and the knowledge is so tightly integrated in the atom that it is integral to the atom. In fact, the atom's virtual energy is always exactly the same as the atom's real energy.
>
> I could keep the universe spread out and localized and I thought about it inasmuch as each new cluster of matter would need to learn how to sustain itself in each and every new space. This is the way to go but I knew new clusters would begin to interact and could become chaotic. So after some local growth new regions can grow only if they become tractable with neighboring regions and some things ended up being scratched. I also drew the line where simple elements could do everything instantly but they had to remain simple – that is, atoms have a finite bound on all of their possible states but each state can be instantly produced. The atom now contains all knowledge it needs locally but anything bigger than that would not be created in its final form. Could not sleep for days."

Somehow I thought you could do anything. Never thought you would get frustrated.

"I can and I did better than just create anything. After the atom, I made all entities without bound. I also have very high expectations of all of my creations. I wanted and gave matter lots of smarts so that the infrastructure can grow and organize on its own, too. After I got the third body going computably, things became much brighter although the challenge still was there and there was no guarantee on how far matter could take it on its own. I was able to forge ahead while scratching less than a fraction of a percent of what was created and I am happy about that. Later on I decided to get some help, you know, which was easy once the capacity for relationships increased dramatically in organic matter."

When you made things without bound, some things can grow and become anything – pushing and shoving and then you have one hell of a universe.

"Knowledge and organization are unbounded as well. The idea is not as much Good versus Bad but good versus better. The best part is that bounds get created by desire or by necessity even though the elements themselves are boundless."

If knowledge is unbounded then memory must be unbounded as well.

"Quite."

And the memory is...

"It is."

Where is it? We talked about virtual energy as knowledge and we also talked about the virtual bank to keep it in. We agreed that virtual is intangible. We know that a book has tangible paper – even ink is tangible. But the work made with words is intangible. So I guess I am asking where is the book and where is the work?

"It is all around us and inside all of us."

Coffee break! Perhaps you visited in the East and it was sufficient there. I can see the cultural aspect reflected in our buildings' architecture and so it is all around us.

"You are doing just great. Run with it."

●●●●●● Genesis

So I guess – I mean, I think that the book and what's written in it is a figure of speech and there is no book as such that would have all knowledge written in it.

"Fine start."

I think that knowledge is available to everyone and so the virtual bank is not a good way of describing knowledge.

"Agree. Bank also deals with exclusive possession, which is the real aspect of duality and as such is not a part of the intangible. Virtual bank was used associatively because banking represents accumulation and safe storage. Sometimes it is OK to use certain associations even though they do not completely cover the situation because you will be able to straighten it out later."

I think that knowledge can be accessed associatively. In fact, knowledge can only be accessed associatively. If I say satellite, I am speaking literally of being high up and seeing things in some detail. Figuratively, though, I speak of a strong and passive bond with a large body. Satellite, then, has a number of linkages of various strengths, each having complimentary and possibly subordinated associations.

"So where is that book?"

In my head, for sure. But the knowledge comes in from .. everywhere. But things I see are tangible and we agree that knowledge is intangible. It can be really confusing to see associations other people do not see. Even if we look at the same things, some people may think it is not important but I think it is important. What gives? Does everyone have his or her own book? If so, which one has the truth?

"You are doing so well I am going to take a nap."

I am? I want answers but so far I got nothing but questions. Unbounded questions at that. Perhaps you should stay awake in case this gets too much for me.

"..."

Everybody cannot be right. So this search for truth can be utterly exhausting, possibly destabilizing. Do I change my mind or do I make the other guy change his. Why cannot he change his mind after an argument? It would be nice if there were a book with nothing but the truth. Now you say there is knowledge everywhere, so there. Is knowledge about differences?

"..."

If truth is about our disagreements then it is also about agreements. Is truth irrelevant, then? Hello?

"You have no idea how well you are doing."

I am going nuts here. Everything means something. As if everybody and everything is talking to me. How come I am the only guy who is not calm? Or is it that the other people are just plain ignorant? Stupid? Maybe they need to get straightened up. Maybe I am the only guy who can see things other people do not see or do not care about. Am I the chosen one? If so, what am I chosen for? What if someone else is chosen for something better? Who has chosen me? Is everybody chosen for something? If so, who is pulling the strings? Who is the puppet master? Yeah, I'm not going to do anything. I'm going to ignore this if I want to. What would free will be good for if there is such a thing as free will! There must be because I am nobody's patsy.

Then again, if I am chosen for something, there may be privileges that come with it. But what if someone claims privileges for something he has no rights to? What gives me the privilege to disclaim his privilege? Our privileges, even if proper, may be in conflict. Are we, can we, should we get in a fight over it? This is not black and white, is it?

"Quite."

So it is about persuasion. It is about influence where I and everybody else can say what they need or want to say. It may be right or wrong or greedy or elegant or anything, while everybody can take it or leave it, for when it comes to knowledge there are no limits. I can accept the information and I can act on the information. Yet the information in and of itself cannot command me to action because information, wherever it comes from, just exists.

● ● ● ● ● ● ● Genesis

"Great. Knowledge is intangible and exists along and together with all other knowledge. As such, knowledge cannot translate into action without the free choice of the one who hears it. As you so keenly observed, virtual energy can be manifested only by converting it to real energy through your person."

Everybody and everything, then, make their own book and everybody and everything can also read everybody else's book. Even inorganic matter has certain abilities and certain knowledge, for atoms can convert virtual energy to real energy and vice versa in accord with their own way of existing without collisions. For the atom, knowledge must exist in space and it is read from there.

"Good. You can do better than that yet."

On the fourth day the Magician created and released the unbounded ethereal memory onto space, and said onto all: "All shall read and add to the history of the ages."

Day Five. Travelin'

When matter becomes a virtual entity it acquires concurrent computational abilities. Since the interactions are logical they can also be nonlocal and distance becomes irrelevant. I am beginning to get the idea how I could build a ship that, in principle, could move logically among the virtual entities and then reduce instantly many light years from here.

> "I think you are pretty close to taking it from here. The knowledge of ages is with you at all times and figuring out the most relevant aspects is now up to you."

> "You may want to consult matter on this, for matter moves in ways that keeps chaos from interfering with the growth of the organized universe. The thing to keep in mind is that matter at times needs to accumulate and at times needs to move. Matter will give you answers that are suitable for matter."

Materializing inside a sun may be a particular solution in a particular context and I can see I would want to pass on that. Right now, superluminal travel is just overwhelming me by the large number of possibilities and in ways things can go wrong. Even if the ship reduces without a problem I may be way off course. What keeps me going is that once I solve a particular intractable problem I will then be in possession of the key and I can reuse that key later on because a particular system solution will not change much. Every time the context increases, though, I have another intractable problem on my hands. Quantum computing will, of course, address intractability but there are no guarantees. How does one know that the computations are on the right track?

> "Keep going..."

There must be influences in the virtual domain that are stronger than others. I think the computations will accommodate the strongest forces of gravitation, momentum, and quantities and distribution of bodies. The solution will then become apparent by the lack of conflicts. I do not see how the reduction would happen by itself. How does the ship then actually trigger the reduction?

●●●●●● **Genesis**

"If you force the reduction you may upset things but you will not break the system. There may be others who may not like what you are doing, though. The universe is organizing and growing and building up. Perhaps you want to be in a place where the ship becomes a needed quantity of matter such as in an orbit of a planet that can use that momentum."

I need to figure out how the logical maps into physical, for if matter is needed close to where I am going, I want to be there logically and then the triggering mechanism can be very small, even natural. What I like is that I will need to know how to organize a system if I want to avail to the benefit. After all this it seems I can travel where matter is needed but not necessarily where I want to go.

"You know the ways of getting there. Perhaps not all the means."

All right. I'll need to transition between real and virtual domains – routinely and at will. While I am in the virtual domain I can think about where I want to go – and I can conceivably get there. Once there I can transition into the real domain if that is how I want to observe the place.

"Enjoy. When you examine your mind and the brain that produces it you will come upon the main thing that makes a human a human. Use it better."

On the fifth day the Magician created the human in the image of both the real and the virtual domains of the universe, and whispered into the human's ear: "A whole person grows and improves without subjugation."

Appendix

1. **Free Will.** Willy woke up one morning and as soon as he looked at the alarm clock he realized he overslept – the clock showing 5:55 am. He called the cab and did not think much of it until the cab pulled in at the curb with number 555 painted squarely in the middle of the door. Rushing the driver on he peeked in his checkbook balance statement. He had an outstanding and amazing balance of exactly $555.55. That was enough for Willy. He told the driver to wait while he withdrew all of it, called in at work to take a day off, and set for the racetrack. He waited patiently and put it all on a horse running number five in the fifth race. Sure enough, the horse finished fifth.

   While Willy technically had a free will, Willy's will was not an active will. Active will always gives persons choices and if anyone claims, "I had to do it, for I had no choice," then the will of the person reduces to that of a machine, for a machine is indeed allowed but one choice.

2. **Formal System.** There is no easy way of getting around the formalism of formal systems. We could say, for example, that now we define a distance of one light-year and since we can add more and more light years without bound the space is infinite. Not so. Spatial distance is unbounded but finite. We can keep on adding more and more light-year increments, but we will never get to infinity because we are still adding, and our children can be adding... The mathematician will probably disagree since it is "obvious" we are going to infinity. The mathematician, however, makes certain assumptions. Once we ask the mathematician to include the non-zero time to actually bridge the locality of space – or take into account the non-zero time of building a more powerful computer – even math will support the inherent finite characteristic of formal systems.

   All real things can at best be unbounded but finite. No real thing can be infinite. If an infinite mathematical series describes compounding or an approach of a real thing then such limit will not reach infinity. If, on the other hand, the infinite mathematical series

Appendix

describes compounding or superposition of virtual elements then such limit will reach infinity.

3. **The Absolute Speed Of Light.** The 1887 Michelson and Morley experiment has shown that the velocity of the light source does not change light's speed or direction. Prior to the experiment, it was presumed that the speed of the light source would affect lightspeed. The Michelson and Morley apparatus was rotated to align one of the light crests to travel perpendicularly to the earth's motion. Presuming that earth's (downward in the illustration) velocity was added to lightspeed, light's crest would travel from **A** to **C** as shown below, and the apparatus would then measure the change in travel distance. Light's travel path **AC** would be longer than **AB** and, therefore, light's crest would take longer to travel back to the detector via route **AC**. The 1887 apparatus was built to resolve the small speed increment the earth's speed was presumed to add to light's speed.

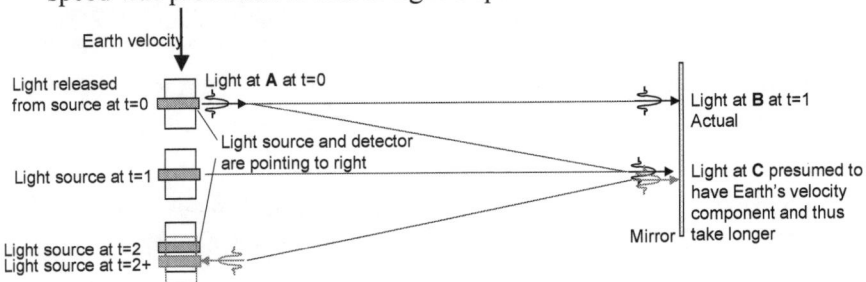

Light Source Downward Velocity Not Added To Lightspeed

However, there is no time difference in light crest arrival no matter how the apparatus is rotated: with, against, perpendicular, or anywhere in between compared to the direction of earth's motion. The correct conclusion is that light's velocity – that is light's speed *and* direction – is *independent* of the velocity of the light source and light always propagates at one constant speed and in direction the light source is pointing at the time of light release. Light's path will always be **AB** and regardless of the velocity of the light source. In the illustration the earth velocity is exaggerated to show that the path **AC** is indeed longer than **AB**. In the actual experiment the difference between the two paths would have been small but the instrument accuracy was good enough to detect and quantify the delay in light's arrival if light were to take the **AC** path. No delay was measured.

As the light source speed is increasing, at some point the returning light will miss the detector – and this is yet another way of confirming that no velocity component is added to lightspeed.

The notion that time appears to slow down with the increase in speed necessitates making a claim that light takes the **AC** path, but light always takes the **AB** path.

The mechanism of light's motion is that light instantly adapts its speed to the medium it encounters. Therefore, lightspeed becomes instantly different in ether, glass, or water. Photons of light also instantly rebound when reflected from a mirror surface and no delay is registered at the turnaround. The medium has priority and light always instantly adapts its speed to the medium it is traversing.

The compact way to state the results of the Michelson and Morley experiment is to say that the velocity (speed and direction) of the light source is not added to light's speed and direction. Another compact way of saying it is that light will propagate at lightspeed in the direction the light source is pointing.

The reason the velocity of the light source cannot be added to lightspeed is because each photon of light propagates independently by interacting with entities photon encounters – in this case the uncommitted electrons of ether. The propagation of light is always logical and not physical; for this reason each photon can make instant turns and slow down or speed up subject to local environment and without regard to inertia. Light is pure energy and light has no inertia. Light source does not and cannot add velocity to light because as soon as light leaves the source, light instantly and logically adjusts to the environment it encounters. If the light source is moving west and the light source releases photons east, photons will travel east at lightspeed. If the light source is moving southwest and the light source releases a photon east, the photon will travel east at lightspeed.

4. **Zero Dimensional.** Mass can exist in three dimensions, but on a cosmic scale mass has been materialized over such large distances that mass bodies became almost zero dimensional points. How large do

Appendix

distances need to be for a mass body to be essentially a point? Heuristically, on the order of a million light years between galaxies, light years between suns, light minutes between planets, and light microseconds between loose atoms.

5. **Getting to Golden Ratio.** The Golden Ratio is often represented as convergence of certain terms from the Fibonacci series of numbers. The Fibonacci series starts with numbers **1, 1**, and each consecutive term is formed by adding the previous two terms. As the Fibonacci series unfolds, the ratios of two consecutive terms converge closer and closer toward the value of $(1 + \sqrt{5})/2$, which is the Golden Ratio and expressed as a ratio of two numbers **a/b**. The Fibonacci series acquired some mystique because of its Golden Ratio convergence property, but it turns out that any numerical series having the next term generated by summing the last two of its terms will also converge toward the Golden Ratio by rationing the two consecutive terms. The mathematical way of saying it is that the convergence is not a function of the initial conditions. The general proof is straightforward. We start with two numbers representing some positive lengths, say **j** and **k**. The next term is, by definition, the sum of the two

$S_0 = j + k$

All partial sums S_n contain numbers **j** and **k**. Next terms are:

$S_1 = S_0 + k$

$S_2 = S_1 + S_0$

...

$S_{n-1} = S_{n-2} + S_{n-3}$

$S_n = S_{n-1} + S_{n-2}$

Now, the S_n/S_{n-1} ratio should converge toward the Golden Ratio **a/b** if our hypothesis is correct. Then, as **n** increases toward infinity

a/b = Limit of S_n/S_{n-1} Substituting for S_n

a/b = Limit of $(S_{n-1} + S_{n-2})/S_{n-1}$ = Limit of $(1 + S_{n-2}/S_{n-1})$

a/b = 1 + Limit of S_{n-2}/S_{n-1}

Quantum Pythagoreans

Expression S_{n-2}/S_{n-1} is a ratio of two consecutive but *inversed* terms and, as **n** increases toward infinity, the ratio converges toward the reciprocal of **a/b**, which is **b/a**. Then,

a/b = 1 + b/a

This equation is not a function of any partial sum S_n and, therefore, **a** or **b** is not a function of **j** and/or **k**. Any value of **j** or **k** can be used in the construction of the series that results in the equation **a/b = 1 + b/a**. Finally, does a solution exist for **a/b**?

a/b = 1 + b/a   now, multiply both sides by **a/b**

(a/b)² = a/b + 1

(a/b)² - a/b - 1 = 0   for positive solution of spatial distance:

a/b = (1 + √5)/2

a = (1 + √5) and b = 2

6. **Just two squares.** Two squares of the same size. One is solid, the other not. Pick the square you like better:

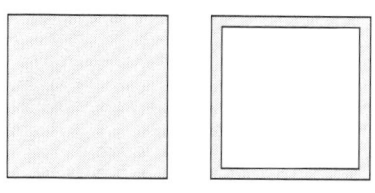

If you were to explain why you picked it, would you change your mind? That is, an explanation with specific meanings is a rational explanation and words such as 'real' and 'practical' will favor the solid square. Irrational descriptions such as 'pretty' or 'intriguing' will favor the empty square. The mechanism is that the solid square has a real – that is truncated – diagonal with fixed and exact distance while the empty space inside the square on the right enables the creation of the infinite superposition of waves that are the root of two. The empty space is the required parameter for the infinite wave formation and if such square were to exist on a tablet, the square's outline would need to rise above the solid part of the tablet or the inside of a square would need to be hollowed out.

Appendix

7. **Golden Ratio Application.** If a bank were to give the customer the amount equivalent to the initial deposit every year, the bank would need to apply the annual multiplier of 1.618 and start paying out the customer after the *second* year.

8. **Applying the *Golden Eye* geometry.** The ancient Alpha & Omega can be constructed. In the construction the alpha becomes the angle as well as the Latin capital letter **A** that take a particular and unique portion of the eye's circle. The *Golden Eye* is likely the best tool for representing the ancient Alpha & Omega and the mystery that encompasses all living things.

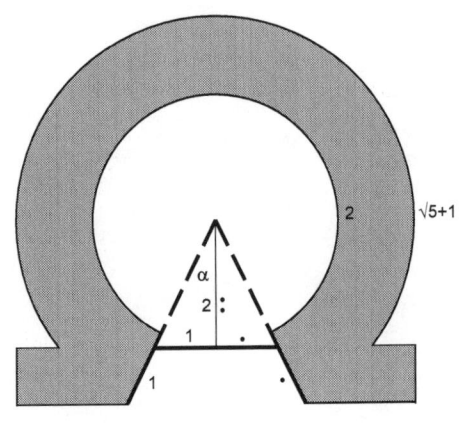

Alpha and Omega through the *Golden Eye*

Note that the sequence of distances is 1, 1, and 2, and these are the starting numbers of the Fibonacci sequence. The ancient Egyptian god of knowledge and magic Thoth is said to have restored $1/64^{th}$ of the Eye of Horus when its owner could not find the missing piece after the battle – but the angle α, which is also prominent in the Great Pyramid, is much larger than $1/64^{th}$ of the full circle. Looking from the bottom up, the accuracy of the pyramid's placement and execution of various angles is much better than $1/64^{th}$ of a circle. For ancient Egyptians the $1/64^{th}$ is the smallest amount associated with the measurement of grain volume and water volume, and could be compared to "less than two penny's worth" today. If Thoth makes a claim to the secrets of the pyramid he should certainly come up with a better story in addition to keeping his arms straight and angled the way he does. Perhaps Thoth thought about the pointlessness of dividing the circle into smaller and smaller sections of a circle. Perhaps Thoth thought about the Infinity gap where the closure of the gap is neither needed nor desired.

There are two things about Thoth of some import. One is his "As above, so below – as below, so above" phrase. The great thing about this phrase is that many, if not most, authors omit the second part "… as below, so above." Such omission tells you nicely that you want to omit altogether what the author is saying. The second thing about Thoth is what is said about the alleged *Book of Thoth* he authored. The spells within this book are said to give the user power over the gods. The truth is the most potent spell.

Not all is said about the Alpha and Omega, however. The Alpha is a specific penetration into the closed circle or sphere. The sphere in general contains wavelengths of all frequencies that are shorter than its diameter, and is thus a repository of infinities. The penetration may release a certain subset of frequencies that may manifest in various forms, and some say this may include visible forms. The Alpha and Omega provides the fundamental geometry behind the ancient *thunderbolt*, the tool of gods.

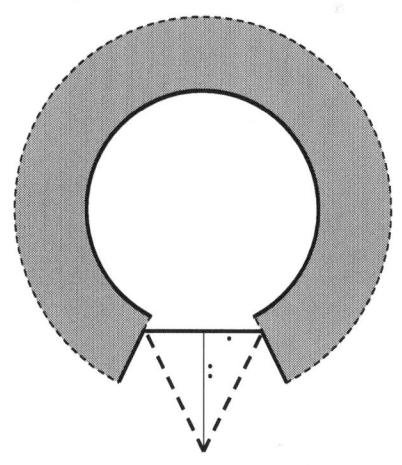

Thunderbolt of the *Golden Eye*

In the case of our construction of the Alpha, the root-of-five frequencies would be preferentially emanating and this particular "thunderbolt" may then have healing properties.

Appendix

9. **Perfect Circle Division by Five.** Division of a circle by geometric means results in exact partitions for some integers. In the text, numbers 2, 3, 5, 15, and 17 are known to divide a circle exactly, while numbers 7, 9, 11, and 13 are known not to be able to divide a circle exactly. The construction of the pentagon and the pentagram belongs among the oldest geometric feats, particularly since the construction is neither obvious nor intuitive.

By using the radius of distance 2, the shortest distance we will need for pentagon construction will be 1.

Instructions:
1. Draw a horizontal line
2. At origin **O** erect a vertical line and draw a circle of radius **2**
3. Make point **A** at distance **1** from **O**
4. Draw an arch around point **A** through **V** and make point **B**
5. Distance **VB** is cord **c** that makes sides of the pentagon

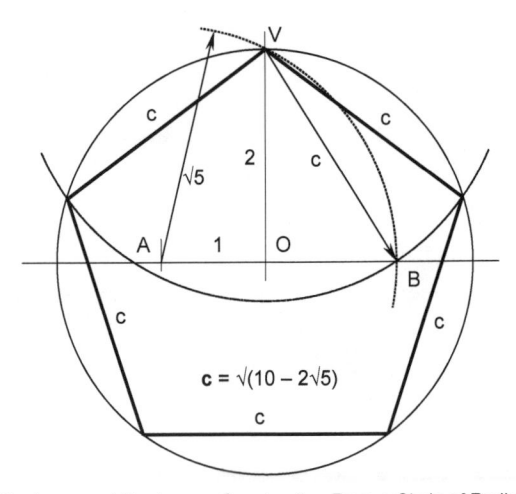

$$c = \sqrt{10 - 2\sqrt{5}}$$

Pentagon and Pentagram Construction From a Circle of Radius 2

The resulting cord length **c** is an irrational number. In the text that introduces the Golden Proportion through the Golden Eye, the pentagon's side is length 2, a rational distance.

10. **Mixed Up.** Taking the output of the inverter and routing it to its input is an allowed combination that has an indeterminate and confusing

Quantum Pythagoreans

outcome. The inverter, as the name suggests, inverts true logic level, say 1, into a false level that is 0, and vice versa. When connected as shown below, the inverter's output cannot make up its mind.

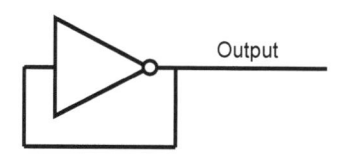

Output: True or False

The reconciliation of the conflict is through enhancement in context, which, in turn, facilitates completeness. Completeness, however, cannot be achieved at the expense of tractability but in this example the outcome is tractable. In the case of the inverter connected as shown above the frequency response of the inverter is included as a new parameter of the formal system to achieve completeness. The delay through the inverter establishes precedence, and the state of the output is then known to any arbitrarily small unit of time.

Of interest may be Gödel's claim that the logic embodied in the circuit above is an example of indeterminacy, which can be mathematically proven. However, for the mathematical proof to hold the delay through the inverter must be zero, which is an artificial construct that does not happen in reality.

In the case of the second law of thermodynamics, the constraints that are made in the laboratory to create a thermally closed system are extended to the entire universe and then claims are made that the universe tends toward entropy – except that the necessity of implementing the closed system construct in the entire universe is now overlooked. It is not possible to isolate the entire universe and the universe is in fact not a closed system.

Also, general relativity calls for observational constraints – the observer is not allowed to look outside the moving laboratory.

All of the erroneous conclusions from these examples are based on artificial and incorrect context construction. Such presumptions may be egregious enough to classify the construction of these contexts as pseudoscience.

Appendix

11. **Some Geometric Relations of the Great Pyramid.** Any proportion of the Golden Numbers **a** and **b** is the Golden Proportion.

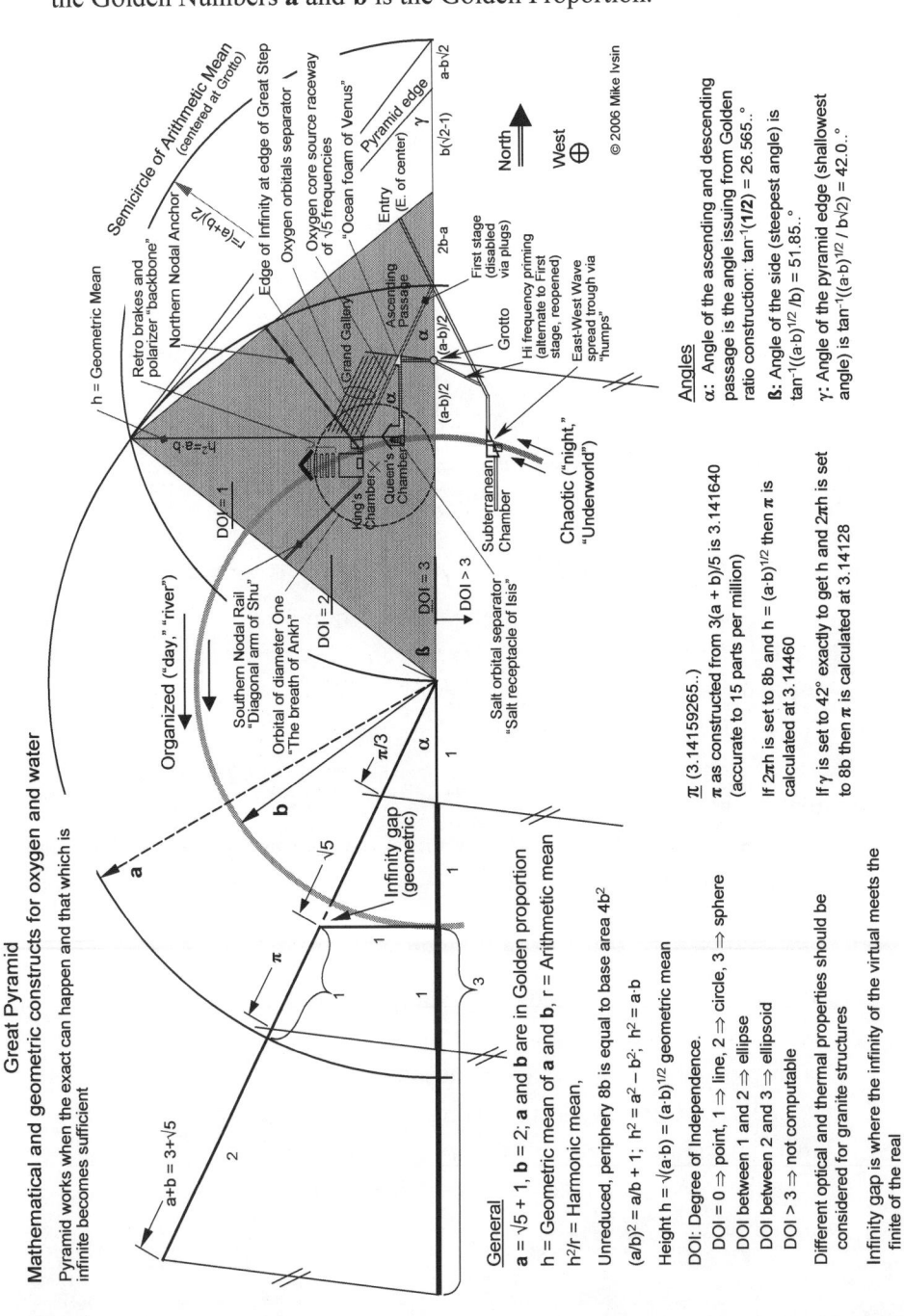

Great Pyramid
Mathematical and geometric constructs for oxygen and water

Pyramid works when the exact can happen and that which is infinite becomes sufficient

General

a = √5 + 1, **b** = 2; **a** and **b** are in Golden proportion
h = Geometric mean of **a** and **b**, r = Arithmetic mean
h^2/r = Harmonic mean,

Unreduced, periphery 8b is equal to base area $4b^2$

$(a/b)^2 = a/b + 1$; $h^2 = a^2 - b^2$; $h^2 = a \cdot b$

Height $h = \sqrt{(a \cdot b)} = (a \cdot b)^{1/2}$ geometric mean

DOI: Degree of Independence.
DOI = 0 ⇒ point, 1 ⇒ line, 2 ⇒ circle, 3 ⇒ sphere
DOI between 1 and 2 ⇒ ellipse
DOI between 2 and 3 ⇒ ellipsoid
DOI > 3 ⇒ not computable

Different optical and thermal properties should be considered for granite structures

Infinity gap is where the infinity of the virtual meets the finite of the real

π (3.14159265..)
π as constructed from 3(a + b)/5 is 3.141640 (accurate to 15 parts per million)
If 2πh is set to 8b and h = $(a \cdot b)^{1/2}$ then π is calculated at 3.14460
If γ is set to 42° exactly to get h and 2πh is set to 8b then π is calculated at 3.14128

Angles

α: Angle of the ascending and descending passage is the angle issuing from Golden ratio construction: $\tan^{-1}(1/2)$ = 26.565..°

β: Angle of the side (steepest angle) is $\tan^{-1}((a \cdot b)^{1/2}/b)$ = 51.85..°

γ: Angle of the pyramid edge (shallowest angle) is $\tan^{-1}((a \cdot b)^{1/2} / b\sqrt{2})$ = 42.0..°

© 2006 Mike Ivsin

Quantum Pythagoreans

It is easy to speculate and difficult to show that the Great and other pyramids were constructed as ways (geometry) and means (stone, energy) of terra-forming the earth. There is a class of life, humans included, which requires a particular form of ecosystem composed primarily of water, oxygen, nitrogen, carbon dioxide, and a multitude of salts. The Great Pyramid is a good candidate as the generator of the water/oxygen component while its operational success is still remembered through the resulting floods. It seems that the water-generating component worked so well the pyramid had to be detuned through the plugs in the ascending passage and also through the fill of the Grotto bypass.

In the Great Pyramid, the vast majority of constructions strive for irrational distances with the purpose of generating a particular family of frequencies. The King's chamber proportions consist of mostly irrational (irreducible) proportions, although one 3, 4, and 5 unit triangle can be found. In the illustration below the units of measure are in units of the smallest integers. Multiply the units by five to obtain Royal cubit measures.

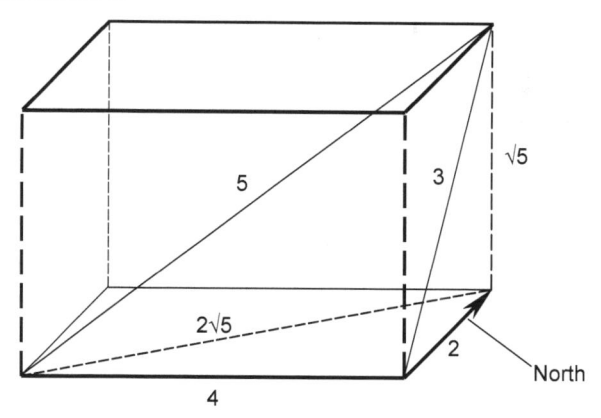

King's Chamber, in fives of Royal cubits

Applying Balmer's $m^2 - n^2$ relation, the result will suffice for the principal orbitals of 2, 3, and 4. For orbital 2 and 4, $\sqrt{5}$ is required and supplied by the trough in the Grand Gallery. In addition to $\sqrt{5}$, the trough also supplies the roots of all square values of $d\sqrt{2}$, where $d\sqrt{2}$ is any distance up to the length of the Grand Gallery.

345

Appendix

One of the first measurement enigmas of the Great Pyramid was the realization that if the pyramid's height **h** is taken as the radius of a circle, the circumference of such circle is very nearly equal to the perimeter of the base – that is, **2πh = 8b**. (**b** is one half of the distance of the base.) Because π is transcendental and has an infinite mantissa, it is not possible to use the stone and construct the base perimeter to be *exactly* the **2π** multiple of its height. Not only that. No matter how close we could construct the distance, such distance is a real distance that does not accommodate *the wave*. By constructing the pinch at the mid-base, however, the circumference that follows the actual stone will be slightly greater than the **2π** multiple, while the circumference going along the periphery in the straight line will be slightly shorter than the **2π** multiple. In this situation the circumference will be sufficiently close to **2π** while the *exact* **2π** multiple is now also available to the waves through the air gap in the pinch. Mathematically, the 'squaring of the circle' is valid at any level of the pyramid and, therefore, the pinch needs to extend all the way to the top. Another way of saying it is that any horizontal cut of the pyramid will have a new value for both its height **h** and half-base **b**, but the **2πh = 8b** relationship continues to hold.

Taking the standard four-sided pyramid's periphery to be the exact **2π** multiple of pyramid's height, the azimuth angle γ of the edge will be $\gamma = \sin^{-1}(\pi^2/8 + 1)^{-\frac{1}{2}}$. The angle value is a constant, which means that the edge of the pyramid is a straight line and, incidentally, very near to the whole of 42 degrees. As a pyramid builder, you would need to settle on a particular real angle at the edge, say, 42.<u>0</u> degrees and then verify that the **2π** multiple of the now slightly different **h** still falls in between the straight-base periphery and the pinched-base (kinked) periphery.

It is easy to apply the Pythagorean theorem and construct any and all irrational numbers except transcendentals. To construct the most important transcendental, which is π, the pyramid geometry is needed. That is, two-dimensional geometry of the Pythagorean theorem is sufficient for irrational numbers that are not transcendental, but for the construction of transcendental numbers we need to go to the three-dimensional geometry. The construction of π calls for a summation of

Quantum Pythagoreans

an infinite number of terms and pyramid geometry allows it. The summation happens concurrently through superposing standing waves and the examination of the $\pi^2/8$ term reveals the particular infinite numerical and *rational* series, discovered by Leonard Euler, which accomplishes this feat.

Should you visit the Great Pyramid, in person or in flight of fancy, bring along the mechanics of the following: the Golden Proportion and its geometric creation mechanism, the exact distance as a solid line (rational distance for the stone), irrational distance as dashed line (empty distance or a gap or a trough), a beam branching mechanism ("splitter"), a musical octave, feet and inches, Tetractys in the form 43210, and the impossible but obvious appreciation that there is more than one path to the center of the maze. You should be able to recognize the edges of infinity and, while the Great Pyramid is no longer operational, you may want to take the radiometer (light mill) with you nonetheless. The past explorers of the pyramid degraded and corrupted the pyramid, both literally and figuratively. Most people do not understand the pyramid and you may want to investigate the trough at the Grand Gallery, for here is where the root of five and roots of all the square diagonals (root-of-two multiplier) were intensely created. If allowed, you may pour a little clean water over the edge of the Great Step. Take note of the square forms running up and on both sides of the trough, for they aid in the Pythagorean construction of irrational distances along the centerline bounded by the trough. Only the diagonals of squares can overlay the centerline inside the trough. The squares forms are in a plane that should be slightly different from the slope of the trough, for the plane of the squares hints at, and superposes with, the true centerline of the root-five standing waves, which are anchored at the edge of the Great Step.

In the antechamber, touch and imagine rotating about the granite stub point issuing from one of the slabs. This point is the "nail" or "fulcrum" that anchors the real and the virtual components together when or if needed. A speculative conclusion is that this anchoring point may be the centering function for existing components rather than having the function of the actual subatomic particle creation. In

347

Appendix

general, a point is the emerging point-of-symmetry for real (male) existence. Circumambulation is wholly a male-enforcement activity.

To make water through computational means, oxygen and hydrogen are made in a particular proportion. In the case of the Great Pyramid both atoms were made *in situ* – that is, everything is made "home grown" from scratch. Atoms as well as the water molecule are made in the virtual domain where much computability and energy in virtual form is brought together and along with particular geometric constructs. Virtual energy needs particular intelligence that is suitable for the construction of real atoms. Symbolically, both Isis and Thoth get involved. Since there are but two chambers in the pyramid and but two atoms are needed for the molecule of water, the commencement of reverse engineering is not that difficult. The oxygen core may be crafted in the King's chamber (sarcophagus being the reduced, or real, core while the rest of the chamber defines the constructs of the pulsing virtual core, which may then suffuse with orbitals). Oxygen's electron orbitals are formed through the upper Grand gallery and the Relieving chambers ("The backbone of Osiris"). But do feel free to form your own conclusions. For example, the hydroxyl radical OH may be the central ingredient. Oxygen may be the most difficult to make but hydroxyls and peroxides are very close to water. Hydroxyls are also highly reactive and could account for the erosion of the King's chamber sarcophagus.

You may call upon the story of Osiris as the story of making oxygen and water. Osiris is confined in a coffin against his will, which is no other than the reduction of the gravitational wavefunction accompanied by the full reduction of the core. The core moves as it is imparted with momentum and Osiris moves confined (in the coffin) down river. Osiris coffin is lodged in a trunk of a tree, which is the backbone (or a post) containing four horizontal partitions. Opposing forces now discover the body of Osiris and his body is chopped up and scattered – that is, the core spreads. The core stops expanding when it suffuses with four electron orbitals. Isis recovers missing pieces and at this point the story should end because the process repeats. But the repetition is true of inanimate matter as well as for things that are earth-bound. Osiris, then, reproduces by producing his image through the virtual domain that is Isis. Because the orbital energy cannot be exactly equal to the linear energy, Isis adds some of her own energy to make the transition a success. You may also include Osiris "props" in the story: The flail and the crook. These are both geometric constructs where the flail is reminiscent of a tetrahedron's axes of projection while the crook is the enticement to the creation of orbits and consequent permanence.

Hydrogen is so simple it may not be necessary to manufacture it. The Queen's chamber may then be for the purpose of making salts, for ions of salt are necessary for most living things. The "power plant" most likely abutted the east side of the pyramid and is now largely, and likely purposefully, decommissioned. However, it is possible that the "power" was generated locally but the *signal* or knowledge or intelligence that was necessary for production of various elements was captured (decoded) through the "funerary" at the east side. All pyramids have highly geometric construction at their east side, while an additional horizontal pointer, or road, leads to this "knowledge decoder." While some think of the pyramid as the crypt, it may well be a crypt as applicable to encryption and decryption.

What is the purpose of all of the other pyramids? These pyramids could be supportive in a sense of creating other inorganic materials such as salts that support life. An intriguing possibility is that each of the other pyramids may support creation of various life forms.

It may seem that the technology allowing effortless trimming and precise shaping of stones in the quarry (sekera, hatchet, or chopper) is sometime in the future. However, the computational knowledge of making matter is applicable to shaping matter. This is the first of the two enabling technologies for the construction of the pyramid. The second is the nullification of gravitation that enables straightforward though still complex pyramid construction. The size of the pyramid stones, then, is not driven by the stones' weight but by the resolution – that is, precision – needed for constructions of particular geometric shapes. Ancient Egyptians did not build the Great Pyramid with indigenous technology.

Unless and until we understand the pyramids' east side decoders, the harnessing of ether borne energies and photonic energies will

Appendix

remain an enigma while most of us will remain earthbound. Yet, we want the technology that we use to be indigenous, for that is the price of independence.

12. **Cellular Automaton.** Because the cellular automaton does not have an objective, there is no intention to create anything at all. A cellular automaton cannot improve itself on its way to a goal because there is no goal. A cellular automaton, given the same initial conditions, will not create a different pattern and will not make any pattern in less time next time around. For given initial conditions, the process will be always the same and therefore the intermediate and the final patterns will be the same. We can, therefore, make a point the other way around. The creation of shapes of real leaves, for example, is a local activity and the leaves, in their growth and shape, interact only with their immediate environment using cellular automaton rules.

A cellular automaton is a local machine that creates very diverse, computable and repeatable 1D, 2D or 3D patterns. The utility of a cellular automaton is based on its memory saving capacity. Without a formula the cellular automaton's growth would be in a series of lookup steps and the upcoming steps through memory fetch would require an extraordinary amount of memory. Taking an example of a typical shape of a leaf of a 1,000-cell by 1,000-cell finite automaton, each step requires a million bits of memory. If the shape of a leaf takes 1,000 steps to develop then the total amount of memory is 10^9 or one billion bits of information. If the leaf runs out of memory after 1,000 steps then the leaf has no choice but to drop dead. The advantage would be that the leaf could develop almost instantly if energy resources were available. Securing gigabits of memory for each leaf just to accomplish the result quickly does not seem to yield much in practical advantage. A cellular automaton, then, is a memory saving device that needs memory only for the initial pattern and one rule that is copied among all cells, which, in turn, can be applied with unlimited number of synchronous steps.

13. Classification of Numbers

Class	Characteristics
Integers	Whole numbers. I~ have no decimal fraction and can be further subdivided into composite, incomposite, and circumpositional families. I~ and Rational numbers (below) together form real numbers. Integers operated on by -1 are also integers but are not included in real numbers. Positive integers are also called Natural numbers.
Rational	R~ numbers result from rationing of integers or exact division of a circle. Decimal fractions always (sooner or later) repeat and, therefore, a decimal fraction is finite.
Irrational or Hyperspatial	Ir~ numbers result from geometric relations among Real numbers in 1D. Ir~ numbers, by definition, have an infinite mantissa, which also means that their decimal fraction never repeats. Positive solutions of finite polynomial equations may result in a representation of Ir~ numbers such as the square root of two. Truncated Ir~ number such as 1.414 is no longer Ir~ number. The exact representation of Ir~ numbers via real numbers is intractable because the generation of an infinite quantity of decimal place numbers is intractable. The actual construction of Ir~ number consists of a straight Ir~ *distance* (vacant separation) between two zero-dimensional points. The representation of an Ir~ number as an actual line (magnitude) facilitates transformation into a real number. In nature, the conversion of a Ir~ number into a Real number is not arbitrary and necessitates termination with a particular fraction, which is at a particular boundary provided by Hyperspatial numbers. Diagonals of squares are Ir~ numbers that can be placed along a given one-dimensional centerline, which constructs a $\sqrt{2}$ multiple of any Real number: a unique and important subset of Ir~ numbers.

Appendix

Class	Characteristics
Transcendental	T~ numbers result from infinite series and exist in 2D. Like Ir~ numbers, T~ numbers have non-repeating and infinite mantissas. Unlike Ir~ numbers, T~ numbers do not result from a root of a polynomial equation. T~ numbers can be constructed only via 3D geometry of specific pyramids, which have intrinsic infinite superposition mechanism. π and **e** are the most prominent T~ numbers.
Harmonics or Quantizing	H~ (or Q~) numbers are specialized subset of R~ numbers. H~ numbers result from a finite subset of an infinite series of decreasing terms. Q~ numbers are a subset of H~ numbers that satisfy quantum mechanical considerations associated with orbitals.
Virtual or Imaginary	V~ entity is in a form *i*b where **b** is a number, variable, or function. V~ number is always double ended, occupies 1D (rare), 2D, or 3D space, and interacts with the environment as a whole. V~ number can transform into a single real number subject to Q~ number termination (boundary).
Other	Any number that does not fit any of the classes above such as the infinitesimal numbers.

Quantum Pythagoreans

14. **Light Mill Reversal Procedure.** When the light mill is exposed to a light source the mill moves such that the dark paddles recede from the light source. Light mill rotation can be reversed in the freezer, as follows:

    a. Prepare a flat area in the freezer to stand the light mill on

    b. Don't leave the freezer door fully open – watch the mill through a small opening. The idea is to surround the mill with cold environment. After about two seconds you will see the dark paddles starting to advance.

    c. Backward rotation becomes faster if you turn off the freezer's light bulb – you will be able to see the paddles okay through the opening and cutting off the radiation from the light bulb adds quite a bit to the reverse rotation. The light bulb's emission provides the mechanism for the forward rotation.

15. **Light's Rotation (Polarization).** In the text, the interferometer photon branching shows that the reflected branch of the photon is moving to the left, as shown below.

    Wavefunction at 50%

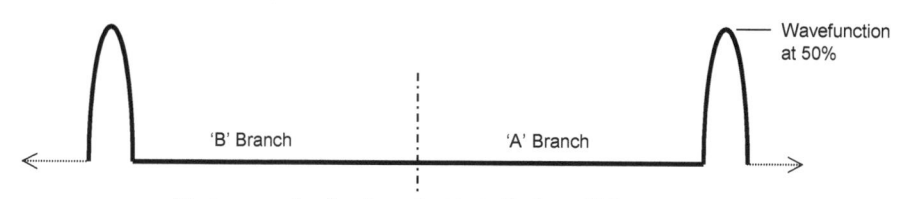

    'B' Branch   'A' Branch

    Photon wavefunction branched by half-silvered Mirror

    Moreover, the experimenter may want to recognize other operations that affect the photon's polarity parameter. As a result of the photon's branching, the reflecting component of the photon wavefunction actually rotates by 90 degrees counterclockwise – while the passing through (transmitted) component does not rotate. In actuality, after the photon is branched by a half-silvered mirror the photon looks as shown below. Note that there is no real (r) and virtual (i) designation on the axes because light operates in the virtual domain exclusively and all wavefunction values carry the i designator implicitly.

Appendix

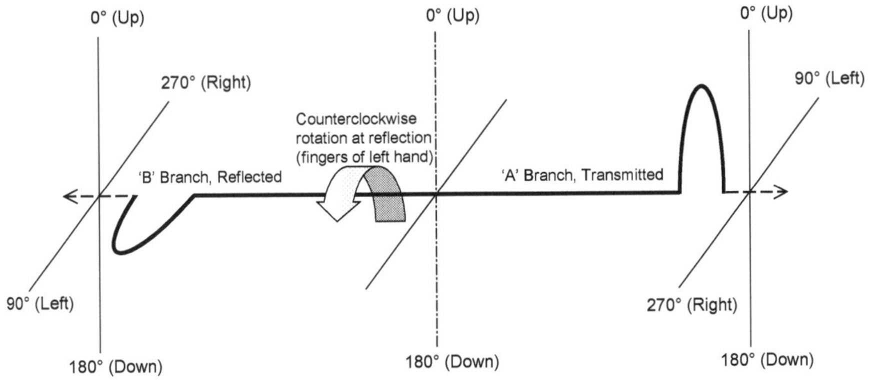

0° (Up)
270° (Right)
90° (Left)
180° (Down)
'B' Branch, Reflected
Counterclockwise
rotation at reflection
(fingers of left hand)
0° (Up)
180° (Down)
'A' Branch, Transmitted
0° (Up)
90° (Left)
270° (Right)
180° (Down)

Photon wavefunction branching by half-silvered mirror: Reflected branch rotates

As discussed in the text the two branches are interconnected. Showing the rotation of the reflected component complicates the explanation of the probability of the photon's reduction, which depends only on the wavefunction percentage that is subject to measurement. The probability of photon's reduction does not depend on polarity – also called polarization – and both components in either branch have the same probability of being reduced because they are 50-50 branches of the original photon.

Furthermore, the reflection of either end via the fully silvered mirror (a regular mirror) facilitates the reflecting component's rotation by full 180 degrees. Rotation by 180 degrees manifests as polarity reversal and this can also be visualized as inversion or a reflection about the axis of propagation. For example, if the photonic wavefunction is pointing *Up* as it is approaching the mirror, it will be pointing *Down* after the reflection. The actual mechanism of polarity reversal is the rotation by 180 degrees counterclockwise. Counterclockwise direction of rotation is always with respect to the direction of propagation. Note, however, that the Up and Down is defined once and irrespective of the direction of propagation. Up and Down is defined *and holds* for both photonic components but Left and Right is *local* to each component. Another way of seeing it is that Up and Down differ by 180 degrees and the direction of rotation – be it clockwise or counterclockwise – does not matter. The Up and Down is the same and stays fixed for either of the two photonic components. The rotation by 90 degrees, however, is where the counterclockwise

Quantum Pythagoreans

direction uniquely manifests as a function of the photonic component's direction of propagation – the rotation resulting in Left polarity is not matched by the 90-degree counterclockwise rotation of the other photonic component. The 90-degree counterclockwise rotation thus becomes unique as Left (or Right) polarization of each component. The Up and Down polarization, however, is defined once and stays the same for both components.

In the experiment that uses only the half-silvered mirror(s) and full-silvered mirror(s), the resulting polarity can only be in one of the four directions: Up, Down, Left, and Right. In general, a photon can rotate by any number of degrees when using specialized filters.

Keeping track of 0 (Up), 90 (Left or Right), and 180 (Down) degree rotations does become important if the branched ends of the photon are brought together to superpose in one location. If a mirror is placed in light's path of either branch then, again, the reflected component rotates 180 degrees. If another beam splitter is placed in light's path of either branch then the reflected component rotates 90 degrees CCW while the passing through component does not rotate. If both ends of the photon are brought together in-phase the wavefunction ends up at 100% at just that location and the photon will always be detected there. If both ends are brought together out-of-phase the wavefunction ends up at 0% at just that location and the photon will not be detected there.

A single photon, regardless of the number of times it is branched, always exists as a superposition of all of its branches. In the illustration below, repeated branching brings a single photon's branches together in-phase and out of phase.

Appendix

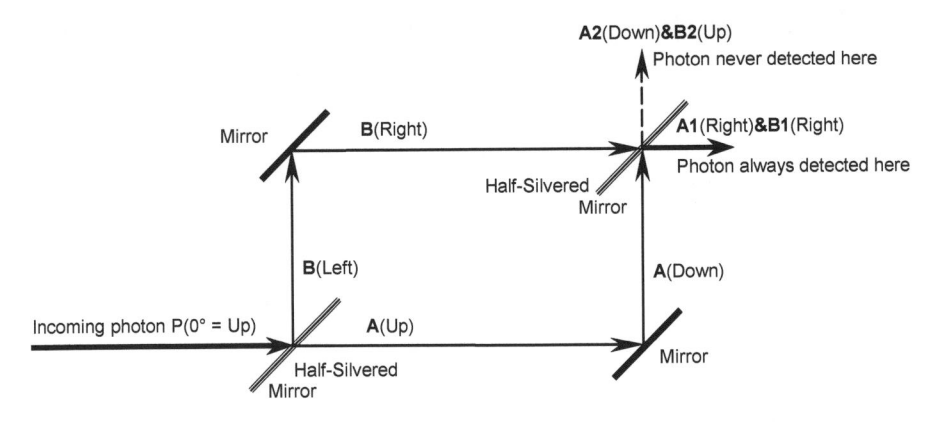

(Figure: Mach-Zehnder-style interferometer diagram)

- Incoming photon P(0° = Up) enters a Half-Silvered Mirror, splitting into A(Up) and B(Left).
- B(Left) travels up to a Mirror, reflecting as B(Right).
- A(Up) continues to another Mirror as A(Down)... *[labels from figure:]* Mirror, B(Right), A2(Down)&B2(Up) "Photon never detected here", Half-Silvered Mirror, A1(Right)&B1(Right) "Photon always detected here", A(Down), Mirror.

P = A + B, A = A1 + A2, B = B1 + B2
Polarity is indicated in parenthesis. All superpositions are self-superpositions.
Good news for the lefties: Point your thumb in the direction the photon is moving (arrow) and your bent fingers indicate the direction of rotation at reflection

Photon branched twice: One path is in-phase and other out-of-phase

If a single photon's components **A** and **B** travel identical distances, components **A2** and **B2** will superpose out of phase while components **A1** and **B1** will superpose in-phase. Since **A2** and **B2** components are propagating in the same direction – rather than just crossing paths – the superposition of **A2** and **B2** will always be zero and the photon (or subsequent photons) will never be detected anywhere along **A2&B2** path but will always be detected along **A1&B1** path. Because all superpositions are self-superpositions, path **A2&B2** wavefunction has zero magnitude, while path **A1&B1** wavefunction has 100% magnitude.

Should you put the proverbial Schrödinger's cat with the poison-releasing single-photon detector anywhere along the **A2&B2** path, the cat will always be alive. However, if any object interrupts the path in any portion of the rectangle, the remaining periphery of the rectangle will become the sole supplier of photons. If the interrupting object is opaque, the object will always attempt to reduce the photon and will do so with 50% success rate. If the opaque object is unsuccessful at reducing the photon's component, the now-full (100% magnitude) photon continues on the only remaining path and splits at the second half-silvered mirror – and then the photon will be able to reduce with equal probability in either the **A1&B1** or the **A2&B2** path. With an

Quantum Pythagoreans

interrupting opaque object, Schrödinger's cat will have a 75% probability of remaining alive with each photon. If the path-interrupting object is a mirror that deflects the photonic component without reducing it, the second photonic component at the other path stays at 50% magnitude and splits equally at the second half-silvered mirror – that is, 25% of the magnitude will be going to the cat's detector. All told, Schrödinger's cat will have 75% probability of remaining alive (with each photon) and regardless of the manner in which the path is interrupted. If one path is blocked and if groups of four photons enter through the **P** path, then, on average with each group: one photon is detected at **A1&B1** path, one photon at **A2&B2** path, and two photons are "lost" because these photons reduce and never reach the second half-silvered mirror. For an important distinction between superposition and self-superposition see Chapter *Self-Superposition*, page 178.

The only time the virtual entity of a photon can be simultaneously in two states is after the photon is branched at the first half-silvered mirror. Two-state simultaneity can then be affected either by self-superposition or by subjecting one photonic branch to a measurement – that is, reduction. When no path is interrupted, the cat will be 0% dead and 100% alive for all photons and for all time. If one path is interrupted then the time duration when the Schrödinger's cat is both 25% dead **and** 75% alive is between the times the photon leaves its source and until the photonic wavefunction component is subject to measurement at the cat's detector.

357

Bibliography

Christoph Riedweg, *Pythagoras*, Cornell University Press, Ithaca, New York, 2005

Robert M. Schoch and R. A. McNally, *Pyramid Quest*, Penguin Group, New York, 2005

Angus Armitage, *Copernicus*, Dover Publications, Mineola, NY, 2004

Peter Aughton, *Newton's Apple*, Weidenfeld & Nicolson, 2004

William J. Boerst, *Isaac Newton: Organizing the Universe*, Morgan Reynolds, 2004

Robert Crease, *The Prism and the Pendulum*, Random House, New York, 2004

Mordechai Feingold, *The Newtonian Moment*, New York Public Library/Oxford University Press, New York/Oxford, 2004

Margaret Jacob and Larry Stewart (editors), *Practical Matter*, Harvard University Press, Cambridge, Massachusetts, 2004

James Gleick, *Isaac Newton*, Pantheon Books, New York, 2003

Joseph LeDoux, *Synaptic Self*, Penguin Books, New York, 2003

David Berlinski, *Newton's Gift*, Simon & Schuster, New York, 2002

Patricia Fara, *Newton*, Columbia University Press, New York, 2002

Mario Livio, *The Golden Ratio,* Broadway Books, New York, 2002

I. Bernard Cohen and George E. Smith, Editors,

> *The Cambridge Companion To Newton*, Cambridge University Press, Cambridge, Massachusetts, 2002

Karl Sabbagh, *The Riemann Hypothesis,* Farrar, Straus, & Giroux, New York, 2002

John S. Rigden, *Hydrogen*, Harvard University Press, Cambridge, Massachusetts, 2002

David H. Clark, *Newton's Tyranny*, W. H. Freeman & Co., New York, 2001

Robin Marantz Henig, *The Monk In The Garden*, Mariner Books, New York, 2001

Quantum Pythagoreans

John D. Barrow, *The Book of Nothing*, Vintage Books, New York, 2000

Joan Dash, *The Longitude Prize*, Farrar Straus and Giroux, New York, 2000

Max Jammer, *Concepts of Force*, General Publishing, Toronto, 2000

Charles Seife, *Zero*, Penguin Books, New York, 2000

Michael White, *Isaac Newton The Last Sorcerer*, Perseus Books, Reading, Massachusetts, 1997

James Burke, *The Day The Universe Changed*, Little Brown & Company, Boston, Massachusetts, 1995

Isaac Newton, *The Principia*, Prometheus Books, Amherst, New York, 1995

G. Harry Stine, *Mind Machines You Can Build*, Top Of The Mountain Publishing, Pinellas Park, FL, 1994

Arthur H. Benade, *Horns, Strings, and Harmony*, Dover Publications, New York, 1992

Mircea Eliade, *Shamanism*, Princeton University Press, Princeton, New Jersey, 1992

Eric Temple Bell, *The Magic of Numbers*, Dover Publishing, Mineola, New York, 1991

John Fauvel, *Let Newton Be!*, Oxford University Press, New York, 1988

Manly P. Hall, *The Secret Teachings of All Ages*, The Philosophical Research Society, Los Angeles, 1988

Rom Harre, Editor, *The Physical Sciences Since Antiquity*, St. Martin Press, New York, 1985

Theodore Cook, *The Curves Of Life*, Dover Publications, New York, 1979

Arthur Koestler, *The Sleepwalkers: A History of Man's Changing Vision Of The Universe*, Penguin Books, London, 1972

G. E. R. Lloyd, *Early Greek Science: From Thales to Aristotle*, W. W. Norton & Company, New York, 1970

Siegfried Giedion, *The Beginnings of Architecture*, Princeton University Press, Princeton, New Jersey, 1964

Sir Isaac Newton, *Optics*, Dover Publishers, New York, 1952 ♪

Made in the USA
Lexington, KY
12 February 2014